AF552720

Managing Soil Health for Maximising Crop Productivity

Managing Soil Health for Maximising Crop Productivity

Editor
Duncan Sones

Authors
**Obeng, F. K., Avornyo, V. K., Fosu, M.
Okorley, E., Yeboah, R. W. N. and Afenyo, E.**

NEW INDIA PUBLISHING AGENCY
New Delhi – 110 034

NEW INDIA PUBLISHING AGENCY
101, Vikas Surya Plaza, CU Block, LSC Market
Pitam Pura, New Delhi 110 034, India
Phone: + 91 (11)27 34 17 17 Fax: + 91(11) 27 34 16 16
Email: info@nipabooks.com
Web: www.nipabooks.com

Feedback at feedbacks@nipabooks.com

© 2017, Authors

ISBN No. 978-93-85516-78-8

All rights reserved, no part of this publication may be reproduced, stored in a retrieval system or transmitted in any form or by any means, electronic, mechanical, photocopying, recording or otherwise without the prior written permission of the publisher or the copyright holder.

This book contains information obtained from authentic and highly regarded sources. Reasonable efforts have been made to publish reliable data and information, but the author/s, editor/s and publisher cannot assume responsibility for the validity of all materials or the consequences of their use. The author/s, editor/s and publisher have attempted to trace and acknowledge the copyright holders of all material reproduced in this publication and apologize to copyright holders if permission and acknowledgements to publish in this form have not been taken. If any copyright material has not been acknowledged please write and let us know so we may rectify it, in subsequent reprints.

Trademark notice: Presentations, logos (the way they are written/presented) in this book are under the trademarks of the publisher and hence, if copied/resembled the copier will be prosecuted under the law.

Composed, Designed & Printed in India

Preface

Soil is the natural medium for the growth of plants and it is doubtful that soilless agriculture will ever be a cost effective alternative for the production of the bulk of the food and fiber needed. For atleast the foreseeable future the world will continue to rely on soil for agriculture. It is therefore essential that the soil is kept healthy.

The present book "Managing Soil Health for Maximising Crop Productivity" has 10 chapters with a bibliography. Its first chapter explores different approaches to manage soil health and soil fertility, Chapter second evaluates the soil fertility and third chapter deals different approaches to manage water in rain-fed and irrigated agricultural systems. Fourth chapter highlights approaches to soil fertility management in general four approaches, namely physical, biological and chemical approaches as well as cropping systems or agronomic approach. Chapter five discusses the concept of integrated soil fertility management (ISFM). There is a outline of the role of cover crops, mulching and conservation tillage as forms of agriculture that can help reduce soil erosion.

Chapter six discusses the use of organic resources in modern agriculture implies the production of crops through organic farming without doing any damage to the environment. Chapter seven and eight deals Mineral Fertilizer Management and Participatory approach to ISFM. The purpose of these chapters are to discuss what a fertilizer is, and the different terminologies used in fertilizer technology. ISFM is important, cost-effective and has the potential of improving farming efficiency and crop productivity. Chapter nine shows what the farmer will go through to determine the viability of the technology and hence be willing to accept it. Farming is a business. It is the income source of the farmer and the source of livelihood and that engages the largest number of people. Chapter ten sharing ISFM information the authors discuss reaching large-scale implementation should be the overriding objective of all project that have successfully ISFM practices to share provide markets support them.

This compilation would be a ready reference and perfect guide to all those in the profession of teaching soil science, agronomy, environmentalists, policy planners, students and the farmers in general.

Authors

Contents

Preface .... *v*

1. Soil Health & Soil Fertility 1

Summary 1

1 The concept of soil health 1

2 What is soil health? 1

3 Soil as an ecosystem 2

4 Micro and macro organisms 2

5 Soil-atmosphere-water continuum 3

6 What does soil fertility mean? 4

7 Types of soil fertility 5

Implications for extension 6

8 Fertility of Soils 6

9 Adjustments to soil 8

10 Causes of soil fertility decline 10

2. Evaluating Soil Fertility **13**

Summary 13

1 Introduction 14

2 Visual diagnosis 14

3 Transect walks 14

4 Nutrient deficiency symptoms 18

5 Common visual symptoms 19

6 Characteristics of plant analysis 24

7 Plant tissue analysis 25

8 Total plant analysis 26

9 Other plant analysis approaches 27

10 Biological tests 28

11 Soil Testing 29

Some Laboratory Equipment for soil and plant analysis 31
12 pH testing 31
13 Yield gap analysis 33
Other Methods 36
14 Resource flow mapping 36
15 Computer-based diagnostic tools 37

3. Water Management 43
Summary 43
1 Mulching 43
2 Bunding 45
3 Terracing 46
4 Contour ploughing 47
5 Irrigation 48
6 Surface irrigation (flood irrigation) 49
7 Localized irrigation 50
Implications for extension 51

4. Approaches to Soil Fertility Management 53
Summary 53
1 Physical approaches to soil fertility management 53
2 Biological approaches to soil fertility 56
3 Use of inoculants 60
4 Cropping system / agronomic approach 62
5 Intercropping 62
6 Crop rotation 63
7 Crop rotation and ISFM 64

5. The Concept of Integrated Soil Fertility Management (ISFM) 67
Summary 67
1 Definition of Integrated Soil Fertility Management (ISFM) 67
2 The context for ISFM 69
3 The merits of ISFM 71
4 Replenishing the pool of soil nutrients 72
5 Adding nitrogen (N) 73
6 Adding phosphorus (P) 74

7 Maximizing on-farm recycling of nutrients 76
8 How soil nutrients are lost to the environment 76
9 The impact of soil erosion on nutrients 78
10 Methods of controlling soil nutrient loss and erosion 78
11 Agronomic/ biological erosion control approach 78
12 Engineering approach 80
13 Improving the efficiency of external inputs 80
14 Protection of beneficial soil micro-organisms 81
15 Integrated pest management 81

6. Organic Resource Management within ISFM 83

Summary 83
1. Organic resource management 83
2 Organic resource quality 84
3 Organic resources available to farmers 85
4 Organic sources of plant nutrient value available to farmers . 85
5 Agro-industrial by-products 85
6 Biosolid 86
7 Processing and application of organic resources 87
9 Fortified compost 90
10 Worms and vermicompost 91
11 Combined application of mineral fertilizer and organic resources 92
12 Crop rotation 93

7. Mineral Fertilizer Management within ISFM 97

Summary 97
1 What is a fertilizer? 97
2 Fertilizer terminology 97
3 Types of fertilizers 98
4 Types of inorganic fertilizers 100
5 Classification of inorganic fertilizers and their Implications on soils and crops 100
6 Management of urea and other nitrogenous fertilizers in agricultural soils 102
7 Management of phosphate fertilizers application 104
8 Which fertilizer should be applied and how? 110

8. Participatory Approach to ISFM Practice 113

Summary 113

1 Why are partnerships important in establishing ISFM programmes? 113

2 Why partnerships are necessary? 114

3 The four phases of partnering 115

Phase 1: Starting up – exploration and consultation (scoping and identifying) 115

Phase 2: Building the partnership (planning, managing and resourcing) 117

Phase 3: Implementing the partnership (implementing, measuring and reviewing) 120

Phase 4: Sustaining, replicating and scaling up (revising, institutionalising and moving on) 122

4 The Participatory Learning and Action (PLA) approach 124

5 Participatory rural appraisal (PRA) 126

9. Economic (Profitability) Considerations of Integrated Soil Fertility Management 131

Summary 131

1 Introduction 131

2 Making a Plan 133

3 Steps to Adoption of a Technology 135

3a Introduction of the technology 135

3b Test the technology 135

3c Farmer's evaluate the technology and compare with the existing ones 136

3d Determine the sustainability of the technology (availability of inputs) 136

3e Try it on a small scale and evaluate again 136

4. Key aspects of the ISFM approach 138

5. Comparing costs of the traditional and new technologies 138

5a How much extra investment or expenditures do I have to make? 139

5b Example: Pepper growing in Mbanayilli in the northern region, Ghana 141

6 Comparing of the tratidional and new technologies 145

6a Calculating input costs through a participatory budgeting process 146
6b Factoring in risk and uncertainty 149
6c Marketing as a risk factor 149
7 Measures of profitability 152
7a Gross margin analysis 152
7b Benefit-cost ratio 153
8 Is it worth doing ISFM? 153

10. Sharing ISFM Information 157

Summary 157
1 Development of a communication strategy 157
2 Development of extension materials 158
3 Creating smallholder-friendly printed material 160
4 Considerations for producing print material for smallholders with low levels of literacy 163
5 Producing farmer-friendly SMS and text messages for extension campaigns 164
6 Producing a dissemination plan to scale-up ISFM approaches 168

Bibliography 175

Colour Plates 179

Chapter 1

Soil Health and Soil Fertility

Summary

Soil fertility is the ability of the soil to support plant growth. The productive capacity of the soil depends on complex interactions between biological, chemical and physical properties of the soil. In essence, soil fertility does not refer only to plant nutrients but also soil organic matter, structure and maintenance of thriving soil organisms. Good farm practices aim at managing the various factors that make up each of these three properties to optimize yields of crops in an environmentally friendly manner. The debate on soil fertility is often erroneously limited to issues relating to plant nutrients.

1 The concept of soil health

It is possible for plants to develop and mature quite well without soil if they are provided with suitable combinations of the five essential environmental factors; light, heat, air, nutrients, and water. This concept is the basis of "hydroponics" and it is not uncommon for plants to be grown commercially in production systems that do not involve soil.

Soil is the natural medium for the growth of plants and it is doubtful that soilless agriculture will ever be a cost effective alternative for the production of the bulk of the food and fiber needed. For at least the foreseeable future the world will continue to rely on soil for agriculture. It is therefore essential that the soil is kept healthy.

2 What is soil health?

"Soil health is the inherent or potential condition of the soil to sustain productivity and maintain a good balance in the environment. Soil health is defined also as the continued capacity of soil to function as a vital living system, by recognizing that it contains biological elements that are key to ecosystem function within land-use boundaries." (Doran and Zeiss, 2000; Karlen and others, 2001).

Definitions of soil health vary. These functions included in the definitions include the soil's ability to sustain biological productivity of soil, maintain the quality of surrounding air and water environments, as well as promoting plant, animal, and human health (Doran and others, 1996).

3 Soil as an ecosystem

An ecosystem is a community of all organisms in a given environment interacting with other organisms and nonliving things in the environment.

Soil is a rich ecosystems, composed of both living and non-living matter with a multitude of interaction between them. Overall, the dynamic processes that occur in soil, create an environment where living organisms are continually affected by changing conditions including seasonal events and climate.

Understanding the role of soil as an ecosystem, and knowing how to manage it in the smallholding, is critical for optimum productivity of soils.

The soil's biological and electrochemical activities cannot be observed directly since they take place at a microscopic and sub-molecular level. Physical changes are however noticeable by a keen observer.

Changes in fertility, tilth and structure may take years to become evident. Early indicators are subtle, and the farmer must be a keen observer to spot them. It is important to recognize the vital role soil organisms play in recycling, releasing, and storing plant nutrients as well as creating better physical environment for plant. Farmers must therefore take decisions that support and enhance the biological life of the soil, which in turn nurtures the crop, maintains soil structure and enhances water availability.

Good tilth is a term referring to soil that has the proper structure and nutrients to grow healthy crops

4 Micro and macro organisms

Billions of organisms inhabit the upper layers of the soil, where they break down dead organic matter, releasing the nutrients necessary for plant growth.

Micro-organisms include bacteria, actinomycetes, algae and fungi. Micro-organisms can be grouped according to their function: free-living decomposers convert organic matter into nutrients for plants and other micro-organisms, rhizosphere organisms are symbiotically associated with the plant roots and free-living nitrogen fixers.

Macro-organisms include earthworms and arthropods such as insects, mites and millipedes. Each group plays a role in the soil ecosystem and can assist in improving the productivity of the soil while sustaining the environment.

The positive benefits of earthworms: Earthworms act as mini-sub-soilers, their burrows increasing soil aeration, drainage and porosity. In the process of burrowing, earthworms mix the subsoil with the topsoil and deposit their nutrient-rich castings on or near the soil surface. The presence of a large earthworm population indicates good soil fertility.

The interaction of the various components of the soil ecosystem also depends on the physical environment. An important physical factor is the soil physical properties that control the fate of water in the ecosystem.

Too much water causes aerobic decomposition to cease and anaerobic bacteria to take over, with damaging effects. For example, nitrification, or the breakdown of nitrate nitrogen to gaseous nitrogen, occurs as a result of anaerobic biological activity in the soil.

Too little water also causes biological activity to slow down that reduces the availability of nutrients.

5 Soil-atmosphere-water continuum

Understanding the fate of water in the soil ecosystem, requires an understanding of the soil-water atmosphere continuum.

a) **The hydrologic cycle** is the term used for the fate of water in the soil. It is influenced by the soil's physical properties. This includes texture, structure and topography. Here the emphasis is on the movement of water between the soil, plants and the atmosphere.

   The effect of these factors on the soil will be examined in chapter 3.

b) **Available soil water** is the available soil moisture that is used by plants and other organisms in the soil ecosystem. The main source of water for the earth is rainfall (sometimes called precipitation). When it rains some of the water is stored in water bodies, depressions on the earth's surface and plant leaves and some percolate through the soil to recharge aquifers. Most importantly, some is stored in the soil as available water.

c) **Transpiration:** Plants take up water from the soil and transport it upward through the roots and stems and release it through the leaves and stems as vapor in the atmosphere (called **transpiration**).

The movement of water through the plant is driven by differences in water potential. Water flows from a high potential to a low potential (imagine free water flow over a sloping surface: water flows from the top, with a high potential energy, to the bottom, with a low potential energy).

In the soil-plant-atmosphere system, the potential is high in the soil and low in the atmosphere. Therefore water moves from soil to plant and to the atmosphere, where it condenses in addition to evaporation from other surfaces and comes down as precipitation.

The movement of water in plants is affected by atmospheric conditions such as **humidity, wind speed, temperature** and **solar radiation**. This movement however, is not only determined by the factors above. It is also affected by the **soil water tension** that is determined by soil physical properties such as texture and **bulk density**.

6 What does soil fertility mean?

Simple definitions

a) Soil fertility is the ability of the soil to support plant growth. It does not refer only to plant nutrients but also soil organic matter, structure and maintenance of thriving soil organisms.

b) Some definitions of soil fertility simply stress the capacity of the soil to supply nutrients. Using this definition, a fertile soil is one that contains an adequate supply of all the nutrients required for the successful production of plant life.

c) Soil fertility can also be defined as the soil's ability to provide essential plant nutrients in adequate amounts and proper proportions to sustain plant growth.

This principle is probably best summed up by the "Law of the Minimum" propounded by Justus von Liebig in the mid-1800's.

Law of the Minimum: This law states that if one of the nutritive elements is deficient or lacking, plant growth will be poor even when all the other elements are abundant. Any deficiency of a nutrient, no matter how small an amount is needed, will hold back plant development. If the deficient element is supplied, growth will be increased up to the point where the supply of that element is no longer the limiting factor. Increasing the supply beyond this point is not helpful, as some other element would then be in a minimum supply and become the limiting factor.

7 Types of soil fertility

The fertility of soil can be considered in different ways, depending on land use and the priorities of the person or group of persons defining the fertility of the soil.

a) The fertility of soil - related to its capacity to support a particular natural community of plants

   In many natural ecosystems, the value of land use may not be clearly defined, and a different definition of soil fertility may be more suitable. In such cases, the plants and other organisms inhabiting the soil in its natural state (e.g. earthworms, bacteria, etc.) are expected to be growing at or near full potential.

b) The fertility of a soil - related to its capacity to maintain consistent output (crop yield) with minimal input.

   Some definitions of soil fertility emphasise the concept of sustainability. A fertile soil in this regard should be able to yield a fairly constant output without any major soil amendment or treatment.

c) The fertility of soil - related to its productivity.

Soil productivity is the ability of the soil to receive and store moisture and nutrients as well as providing a desirable environment for all plant root functions, resulting in optimum yields.

In intensively managed agricultural and horticultural systems, and even in forestry, soil fertility can be defined in terms of the value of products produced relevant to inputs used (including economic aspects of nutrient budgeting). Alternatively, the emphasis may be on quality or **productivity**. A nutrient rich soil is not necessarily a productive one. To be productive, soil must also provide a satisfactory environment for plant growth and the nutrients it contains must be available for use by the plants. It is important therefore to take a broader view of soil fertility and see it as a complex of soil chemical, physical and biological factors that affect land potential and the degree to which a soil is productive.

Fertile soil is characterized by ongoing, complex interactions involving organic matter, macroorganisms, and microorganisms to make inorganic mineral ions available in the soil. Roots absorb these mineral ions if they are readily available in water (soil solution) and not 'tied up' by other elements or by alkaline or acidic soils. E.g. Phosphorus (P) is widely reported to be deficient in Ghanaian soils, especially northern Ghana because of its fixation by aluminium (Al) and iron (Fe).

Soil microbes play a critical role in ion uptake and in the cycles that permit nutrients to flow from the soil to the plant. The microbiological community of the soil system around the roots breaks down the available organic material in the soil into a usable form that the plant root system can readily absorb.

When soil fertility is considered in terms of the highest practical level of productivity, the focus is mostly on physical and chemical aspects of soil. It is important to note that

Some aspects of the biological component of soil fertility can be overridden by addition of fertilizers, but this is not a simple phenomenon, because increased plant growth associated with addition of fertilizer can increase other aspects of the biological activity in soil.

The presence of micro and macro nutrient elements is not enough to ensure adequate growth of plants. Other important factors include the availability of water, soil temperature, and light. Soil physical factors such as texture and structure affect water availability and root penetration. Adequate aeration is another factor that is important for the growth of plant.

Factors inhibiting adequate crop growth even in the presence of nutrients.

Factors that inhibit adequate growth of plants are important characteristic of productive soil. These growth-limiting factors include:

- Sodicity
- Acidity (See chapter 2, section 12 on pH)
- Alkalinity
- Salinity

Implications for extension

8 Fertility of Soils

Soils in Ghana vary in their physical and chemical properties mainly due to the influence of climate and vegetation, including other organisms, which act upon the parent materials.

These soils are further modified by local topography over a period of time. Thus, in each of the agroecological zone of Ghana have some typical major soil groups. Generally, most soils in the various ecological zones of Ghana have low fertility either due to inherent infertility or human induced infertility. Figure 1 is a map of Ghana showing the various agro-ecological zones.

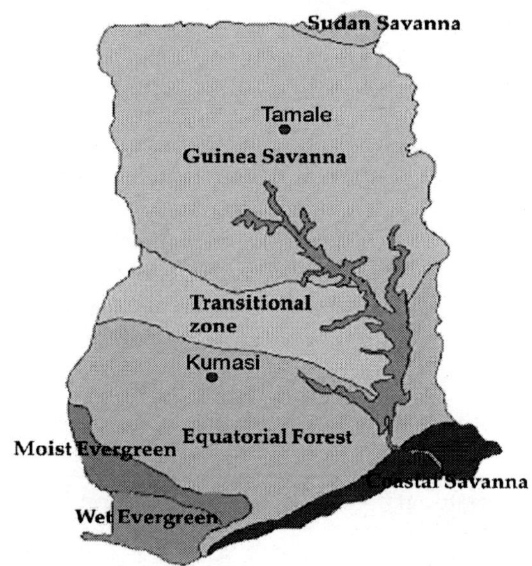

Fig. 1. Map of Ghana, Showing the various agro-ecological zones

The agroecological zones of Ghana and their associated major soil groups

Agro ecological zones.	Soil characteristics
High Rain forest Zone (HRFZ) (Evergreen, and moist evergreen)	**Annual rainfall:** High rainfall of 1,800 - 2,000mm per annum has resulted in highly leached and acidic soils with low base saturation. High accumulation of organic matter in their surface horizons due to the greater accumulation of biomass. Blocking nutrient flows leaving the farm ('leaks in the system')by controlling erosion, run-off and leaching is important in this zone.
Semi Deciduous Forest Zone (SDFZ) (Equitorial forest)	**Annual rainfall:** Ranges from 1,400mm -1,700mm results in similar conditions as in the High Rain forest Zone with less leaching and acidic soils – but not to the same extent.
Interior Savanna Zone (Sudan and Guinea) (ISZ)	**Annual rainfall:** Varies from 900-1,200mm, there is less leaching and therefore, the soils are relatively less acidic. The interior Savanna has only

	one rainfall season and the soils are dominated by concretional soils often forming iron pans at a depth of less than hundred centimeters below the surface. E.g. Kpelsawgu and Changnalili series (Plinthosols). In the Interior Savanna Zone, however, the nature of the soil's parent materials mostly sandstones, shales and the slow accumulation of organic matter has resulted in low fertility status of the soil relative to those of the Semi Deciduous Forest Zone.
The Coastal Savanna Zone (CSZ)	The problems are similar to those of the interior Savanna. The Coastal Savanna Zone has the additional problem of salinity associated with the major soil groups bordering the saline coastal lagoons and creeks. The salt affected soils. The soils have very low reserves of organic matter, nitrogen and phosphorous.
Forest-Savanna Transitional Zone (F-STZ) (Transitional zone)	

Biomass is biological material derived from living, or recently living organisms.

9 Adjustments to soil

Adding nutrients to replenish stocks and flows in the soil	Crops remove significant amount of nutrients from the soil and it is important that these nutrients are replaced after crop removal or harvest. Soils that have low fertility prior to cropping can have their fertility status improved. Inorganic fertilizer that contains the needed nutrients such as nitrogen, phosphorous and potassium can be applied to improving fertility. Organic fertilizer (see chapter in the form of manures, compost, crop residue can be used in areas such as the Savanna zone where the organic matter content of the soils is usually low. Cropping systems that take advantage of nitrogen fixation such as cropping systems that involve growing of legumes can be used to add nitrogen to the soil.
Blocking nutrient flows leaving the farm ('leaks in the system') by controlling erosion,	Nutrient losses can be reduced by controlling: **Erosion losses** can be reduced through the construction of bunds, terraces, or stone lines. **Leaching:** Trees can be instrumental in

run-off and leaching.	recapturing nutrients leached from the subsoil while green manures capture nutrient which could otherwise leach beyond reach of crops into their leaves which are ploughs back into the top soil. Soil treatments such as addition of organic manure or organic matter increase the cation exchange capacity of soils and also increase the soil's ability to hold nutrient against leaching. This strategy can be useful in the high rain forest zone where leaching is a problem.
Optimise nutrient recycling	**Nutrient cycling**, or integration of crop and livestock management, within the farm means: household waste composted organic matter bedding in stables which absorbs urine and conserves nutrients (farmyard manure) and crop residues are incorporated into the soil
Improving the efficiency of nutrient uptake	**Nutrient recovery** may be enhanced in several ways: It is important that nutrient addition be synchronized with plant demand for nutrients. **Fertilizer application** may help greatly. Applying fertilizer close to the roots helps ensure plants take up nutrients more efficiently. **Better agronomic practices** like controlling weeds, pests and diseases should greatly increase nutrient recovery from fertilizers. **Mulching** with rice straw or other plant residues (or sometimes plastic) should conserve moisture and smother out weeds, enabling better crop establishment and nutrient uptake. **Balanced fertilization** is also an important consideration. When nutrient supply is unbalanced, yields and profits decline and, quite often, the quality of the crop is impaired.

Implications for extension: A careful diagnosis of fertility problems for every farm, or portions of a farm, will really help smallholders with the management of fertility on agricultural lands for optimum productivity. However, this resource intensive.

10 Causes of soil fertility decline

Soils in Ghana have low fertility and needs to be improved to make them more productive. There are two main causes of the low fertility level associated with the Ghanaian soils.

a) Natural causes of low soil fertility - soil formation

The soil undergoes many changes with time. Soil formation has a very long history that spans from thousands to millions of years. The soil profile is continually changing and developing with time. Additionally, with time, the climate promotes changes along the earth surface and soil profile development.

From the moment soil is formed, nutrients are being continually removed from and added. The conditions that are present during soil formation ultimately determine how much and what kind of nutrients the soil can naturally supply and hold.

Factors influencing soil formation

- **Finely textured parent materials** tend to weather into finely textured soils that have higher water holding capacities. Most of these soils, however, have low cation exchange capacity hence nutrient retention capacities.
- **Coarsely textured parent materials** end to weather into coarsely textured soils that have low water retention capacity. Infiltration is therefore very high in these soils leading to leaching of nutrients.
- **Grassland soil** will be remarkably different than soils that develop in forests in their organic matter contents. Grassland soils generally have lower organic matter. This soil formation factor is intimately influenced by climate and changes over time.
- **Climate's** effect on soil formation cannot be underestimated. Climate is a fundamental force of weathering that interacts with all other soil formation factors. The primary forces of climate are rainfall and temperature. While nutrients can be released from primary minerals during weathering, high levels of rainfall and temperature that are characteristic of the Ghanaian climate can remove, or leach, nutrients from the soil, and reduce its fertility.
- **Rainfall** intensity greatly influences the extent of soil weathering. Soils that are in a predominantly moist environment tend to be highly weathered. The occurrence of leaching, or the movement of minerals through the soil profile with water, increases as the amount

of moisture increases. In the tropics rainfall is a fundamental force that weathers many soils.

- **Living organisms** including the vegetation that covers the soil, the macroorganisms that live upon the soil, and the soil microorganisms that inhabit the soil have profound effect on soil fertility.
- **Topography** describes the relative steepness of a slope or the flatness of a plane. It may also describe the orientation of the land with respect to the direction of the sun, or it may describe depressions and elevations of the land.Topography has a direct impact on water drainage, depth of soil and nutrient availability. Soils formed along slopes are usually very shallow due to the loss of the soil to erosion. Valley bottom soils are usually very deep and full of organic matter washed down from upland. This is one reason why organic matter content of some soils in Ghana is as low as one per cent whereas other places in the same ecological zone have as high as 3%.

b) Anthropogenic (man-made) causes of low soil fertility

In addition inherent soil infertility in Ghana, human activities such as the following have led to soil fertility decline:

- Land preparation practice & slash and burn, stripping organic matter from the field
- Poor management of crop residues
- Over cropping of a land area without any nutrient management
- Poor fertilizer application and
- Overgrazing
- Burning

These practices, and many others, may cause loss of soil fertility through erosion, loss of soil nutrient and the soil's ability to hold nutrients, breakdown of the soil physical properties and loss of the soil's ability to hold water.

Soil therefore loses the ability to sustain crop production and the natural soil ecosystem.

Implications for extension: Extension activity that seeks to improve crop production, or enhance livelihoods, without addressing the man-made causes of low soil fertility will not create sustainable change.

Chapter 2

Evaluating Soil Fertility

Summary

The soil is the basis of agriculture as the productivity of the farm enterprise is strongly linked to the fertility status of the soil. The soil acts as a reservoir for the essential nutrients that crops need for growth. When these nutrients are in sufficient quantities and balance, agricultural productivity is greatly enhanced. On the other hand if the nutrients are in short supply and imbalanced, crop growth is hampered. When crops grow and produce harvest, they remove nutrients from the soil. These nutrients are exported when the grains or edible parts are eaten and the stover removed. Consequently with time, a soil that is considered fertility will also lose its fertility if no nutrients are added periodically. It is important to assess the fertility status of the soil before planting and all necessary ameliorations effected in time to ensure a good harvest. Various methods exist for evaluating soil fertility. The best indicator of a fertile soil is good growth of crops growing on it. To evaluate the fertility of soil in an area, a quick transect walk can be undertaken to record all observable features and nature of crops growing on it. Soil fertility can also be evaluated by observing nutrient deficiency symptoms. A lack of specific plant nutrient is indicated by typical identifiable coloration of the leaves or tissues of the plant. This approach is post mortem and may lead to reduced yield even with amelioration. Plant analysis where the whole plant or portions of it is sampled and analyzed in the laboratory or on the field, is also a key method of soil fertility evaluation. The assumption here is that a sufficient amount of the nutrient in the plant is an indication of its sufficiency in the soil. The most authentic and direct approach however, is testing the soil itself in the laboratory. Here the soil has to be sampled carefully and analyzed in the laboratory for the concentration of the nutrients of interest. The nutrient levels so determined can be assessed directly in relation to amounts required for a good crop growth. The data can also be used with computer-based diagnostic tools that are usually crop models such as NUTMON, NuMaSS, DSSAT or APSIM to simulate crop growth and development.

1 Introduction

Soils are important for plant growth and development. In agrarian societies such as Ghana, the quality of life and wellbeing of large portions of society are often determined by the fertility of the soil.

The understanding of good farming begins with understanding of the soil. When the soil is poor, the people will be poor. Soil is fundamentally important.

There is a great variation in the length of time soils can be cropped before yield reduction occurs. When smallholders, with generally fertile soil, experience a decline in yield it is an indication that either nutrients or some other component of soil fertility are being compromised.

Soil fertility evaluation is a measure or assessment of the ability of the soil to sustain optimum production of a particular crop.

Diagnostic tests of soils are used for trouble-shooting or as a preventive measure against low yields. Tests include:

- Visual diagnosis - observation of plants in the field
- Obtaining information on past management
- Soil tests
- Tests on the plants growing on it.

2 Visual diagnosis

Simple visual diagnosis is a method of evaluation based on the knowledge of soils and native plants associated with rich or poor soils. It requires some indigenous knowledge about vegetation and soil types.

Here diagnosis is done partly by associating occurrence of some particular plant species on a field with high or low soil fertility. Soil colour, texture, presence or absence of earthworm casts, and other soil attributes may also be used to evaluate fertility of soil based on expert knowledge.

Critical observation is required in this approach.

3 Transect walks

On a transect walk, the agriculturist or farmer walks from one end of the field to the other. Along the walk they note the cause and effect relationships between different landscapes/ topography, soils, resources, cultivation and natural vegetation, land use including human settlement, earthworm casts, and other attributes that indicate low or high soil fertility.

This technique is simple, easily adopted and replicated at the community level. A transect walk usually takes a couple of hours to complete. It requires a good facilitator (to ensure that a full discussion occurs with local analysts) and analysts with a good knowledge of the area, cropping systems, cultural practices and customs and agronomic practices.

The precision of this method of evaluation is limited as it depends on the experience of the people undertaking the transect walk and to some extent the characteristics present at the time. However, it is a useful and quick method of evaluation that gives some indication of the degree of fertility.

When a transect walk is performed as a participatory practice it can:

- Identify major problems and possibilities perceived by different groups of local analysts in relation to the transect
- Clarify local technology and practices
- Corroborate data collected through other tools

How to do a transect walk and produce a transect diagram

Step 1 : Select local analysts

Identify the groups of people (analysts) to talk to about their perceptions of their community and its resources. These decisions will be based on the objectives and depth of information required for the research. For example, separate groups of men and women might be useful because women and men might use different resources:

- **Women** will probably show the resources and features they think are important (such as water sources, firewood sources, and so on)
- **Men** will probably show those they think are important (such as grazing land, infrastructure, and so on).

Other ways to breakdown the group could be class, ethnicity or cultural groupings. A group of five to ten local analysts should reflect any relevant and important social divisions.

Step 2 : Provide introductions and explanations

The facilitator and note-taker introduce themselves and explaining the objectives of the walk and lead a discussion to ensure the local analysts understand, and feel comfortable with, the approach.

Step 3 : Do a transect walk and produce a transect diagram

Discuss with the local analysts the route they would like to follow on the walk. This decision could be based on the community resource map (if one has already been produced). Ask the analysts to think carefully and plan a route that covers the main variations in topography and other features they want to see and show during the walk. Explain that the route does not have to be straight, but can meander if necessary.

With the local analysts, start at the edge of the area and begin the walk. As the walk progresses they stop at key features or borders of a new zone (such as residential, topographic, land usage) and record the distance from the last zone. As an alternative, stop every 100 paces and record.

The local analysts are encouraged to discuss and describe everything encountered or noticed and to explain the key characteristics of areas/ features they see. This discussion can be facilitated by asking questions about the details and by making observations.

The note-taker will record the details that the local analysts encounter-drawing sketches where necessary.

It is not necessary to stick to the original planned route. Deviate when useful or interesting, or even at random, to observe the surrounding area and to gather relevant and useful information.

The walk should be slow to allow different local analysts and try to understand and/or explain the physical features in the village from their perspectives. People encountered along the way should also be interviewed to obtain local perspectives from people who might not have been able, or felt able, to join the original local analysts (these interviews might provide interesting perspectives from people usually marginalized during formal activities).

After the transect walk has finished, sit down with the local analysts to discuss and record the information and data collected. Prepare an illustrative diagram of the transect walk using the information. Where more than one transect walk has been completed, prepare a combined chart and compare results.

The diagrams can be prepared on a large sheet of paper (or on the ground). On the top line, illustrate the different zones that the local analysts visited. Down the side, list headings of the areas of interest (plants, land use, problems, drainage system, and so on) and then fill in the details of what was observed in each zone.

Step 4 : Analyze a transect diagram

It might be useful to have a list of key questions to guide a discussion about the information gathered during the transect walk.

Key questions might include

- What resources are abundant or scarce?
- How do these resources change through the area?
- Which resources have the most problems?
- Where do people obtain water and firewood?
- Where do livestock graze?
- What constraints or problems are in the different areas?
- What possibilities or opportunities are in the different areas?
- How will a proposed policy change or implementation affect the features and characteristics of different areas?
- Where do different population sub-groups live? Are they segregated or mixed?
- Do the poorest households live in certain areas (such as on the edge of an area/community)?

If local analysts have sufficient time, it might be useful to ask them to draw a series of diagrams to illustrate changes over time. If there are several different groups, ask each group to present its diagram to the others for their reactions and comments. Are there serious disagreements? If so, note these and if a consensus is or is not reached.

Step 5 : Conclude the activity

Check again that the analysts know how the information will be used. Ask the analysts to reflect on the advantages, disadvantages, and the analytical potential of the tool. Thank the local analysts for their time and effort.

Points to Remember

Good facilitation skills are key. The approach outlined above is a general guide; be flexible and adapt the tool and approach to local contexts and needs. Large sheets of paper and markers are required

If the diagram is drawn on the ground, then a large area will be needed, as well as a range of objects such as sticks, stones, leaves, seeds, and so on that the analysts can use to represent features on the diagram.

4 Nutrient deficiency symptoms

The nutrient requirement of plants depends on many factors including:

- Specific plant species and variety
- Yield level required (harvest and markets)
- Environment (i.e. soil, water, temperature, solar radiation)
- Management

When nutrient requirement of a plant is not fully met, the insufficiency or limitation of the particular nutrient will be expressed in various forms that can be observed visually.

Nutrient deficiency symptoms can be observed as an abnormality in the colour of the leaves or the growth of the plant. The changes in colour or growth are peculiar for each nutrient element. This makes it possible to identify which particular nutrient is limiting. However, there are multiple other causes of some of the visual symptoms - so caution is required.

Other causes of visual symptoms that need to be eliminated

- Rainfall patterns since planting - drought will cause browning of the leave edges
- Recent applications of weedicides/herbicides
- Pest problems
- Soil preparation
- Fertilizer applications and other soil treatments
- These can also create changes to plant and leaf characteristics.

Limitations of relying on visual symptoms

- **They may be caused by several nutrients deficienciesExample:** Nitrogen (N) deficiency symptoms may be identified, although sulphur (S) may also be deficient and its symptoms may not be readily apparent.
- **Deficiency of one nutrient may be related to an excessive quantity of anotherExample***:* Large quantities of iron (Fe) may induce manganese (Mn) deficiency. Also, plants grown in a low phosphorus (P) soil may not require as much nitrogen (N) as with a soil with adequate phosphorus (P). In other words, once the first limiting factor is eliminated, the second limiting factor will appear (Liebig's law of the minimum see Chapter 1 - section 6b).
- **Other plant stresses (diseases, insects, herbicide damage, etc.) can be difficult to distinguish from nutrient deficiencies**.
- **Visual symptoms may be caused by more than one factor**.
- **Example***:* Purpling may also be seen in some cereal varieties that are rich in anthocyanins. This accumulation may be caused by low phosphorus (P) availability, low soil temperature, insect damage to the roots, or nitrogen (N) deficiency. Phosphorus (P) deficiency often presents as stunting in cereals and the formation of small leaves in legume crops.
- **Nutrient deficiency symptoms appear too late to correct the deficiency without yield loss** – so in most cases they can only:

  Nutrient toxicities even some non-nutrients like aluminum (Al) - can produce visual symptoms similar to nutrient deficiencies. For example, manganese (Mn) toxicity symptoms are similar to iron (Fe) deficiency. Aluminum (Al) toxicity looks like calcium (Ca) deficiency

Anthocyanins are water-soluble vacuolar pigments that may appear red, purple, or blue depending on the pH.

Chlorosis is a condition in which leaves produce insufficient chlorophyll. As chlorophyll is responsible for the green colour of leaves, chlorotic leaves are pale, yellow, or yellow-white.

5 Common visual symptoms

a) **Stunted growth** is a common symptom of all nutrient deficiencies and many other plant stresses. The appearance and degree of stunting may vary with the deficiency. Shorter internodes and a stocky plant or thin, spindly leaves and stem may result in stunting.

b) **Chlorosis or yellowing** is the most common nutrient deficiency symptom. There are many factors causing chlorosis and many forms can be observed. Chlorosis generally results in the yellowing of the entire leaves in a fairly uniform manner. Nitrogen and sometimes potassium deficiency can cause this type of chlorosis in maize.

Chlorosis: Nitrogen (N) deficiency in maize

c) **Interveinal chlorosis:** In this type of chlorosis, the leaf veins remain green while the rest of the lamina or tissue between the veins turns yellow. This type of chlorosis is often caused by iron (Fe) deficiency in maize.

Interveinal chlorosis: Iron (Fe) deficiency in maize

d) **Necrosis or firing** results in the death of plant tissue, although chlorosis usually is observed first. Necrosis often begins on tips or edges of older leaves. Some parts of the leaf may be alive while some parts especially the tips and margins may die and turn brown. Severe nitrogen (N) deficiency may cause necrosis. Damage from drought, herbicides, disease and salts also cause firing or necrosis.

Necrosis or firing in maize

e) **Abnormal colouring** is associated with specific nutrient deficiencies. The observed colours vary from red, purple, yellow, brown or abnormally dark green colouration.

Reddish purple colouration is due to a purple pigment resulting from a build-up of plant sugars when normal translocation to roots, grains, etc. is restricted.

Purple colouration of lower leaves is caused by phosphorus (P) deficiency.

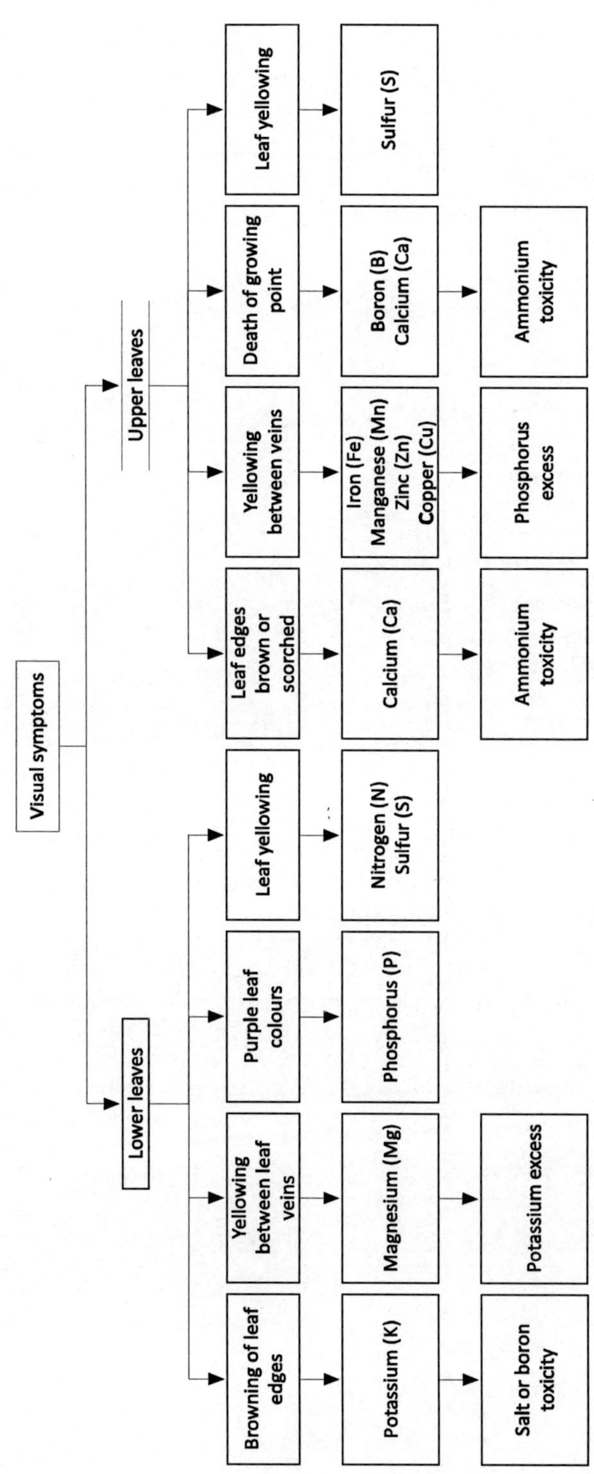

Visual symptoms
Lower leaves
Browning of leaf edges
Potassium (K)
Salt or boron toxicity
Yellowing between leaf veins
Magnesium (Mg)
Potassium excess
Purple leaf colours
Phosphorus (P)
Leaf yellowing
Nitrogen (N)
Sulfur (S)
Upper leaves
Leaf edges brown or scorched
Calcium (Ca)
Ammonium toxicity
Yellowing between veins
Iron (Fe)
Manganese (Mn)
Zinc (Zn)
Copper (Cu)
Phosphorus excess
Death of growing point
Boron (B)
Calcium (Ca)
Ammonium toxicity
Leaf yellowing
Sulfur (S)

The location of the deficiency symptom is related to the mobility of the nutrient in the plant.

Mobile nutrients	Immobile nutrients
The deficiency symptom occur in the older leaves first and will extend to the younger leaves only when the deficiency becomes severe as the plant grows older.	When the nutrient is immobile in the plant, the deficiency symptoms will occur in the younger leaves first.
Examples of mobile nutrients are:	Examples of immobile nutrients are:
Nitrogen (N)	Iron (Fe)
Phosphorus (P)	Boron (B)
Potassium (K)	Manganese (Mn)
Magnesium (Mg)	Copper (Cu)
Zinc (Zn)	Calcium (Ca)
	Chlorine (Cl)
	Molybdenum (Mo)
	Sulphur (S)

General nutrient deficiency symptoms (Mobile Nutrients)

Element	Leaf coloration	Crop growth	Impact on yield
Nitrogen (N)	Light green or yellow on lower leaves which eventually die (necrosis).	Plants stunted and weak; early leaf loss; small leaves; plants mature earlier.	Decrease in yield.
Phosphorus (P)	Purple or dark blue-green; first appears on lower, older leaves.	Plants develop and mature slowly.	Decrease in yield (reduction in tillering in rice).
Potassium (K)	Yellow, brown tissue between veins, necrosis along leaf blade; symptoms on mature, lower leaves; chlorosis or white 'specks' on leaves (legumes).	Slow growth; shortened internodes, leaves can appear wilted.	Decrease in yield.
Magnesium (Mg)	Chlorosis beginning at center and edges on leaf.	Slow growth; delayed.	Decrease in yield.

	blade of mature, lower leaves.	maturity, poor tillering.	
Zinc (Zn)	Yellow or white coloration between midribs and leaf margin; appears on older, mature leaves.	Shortened internodes; rosetting in broad leaves.	Decrease in yield.

General nutrient deficiency symptoms (Immobile Nutrients)

Element	Leaf coloration	Crop growth	Impact on yield
Iron (Fe)	Chlorosis between leaf veins; yellow or white young leaves.	Reduced plant growth or plant death.	Decrease in yield.
Boron (B)	Yellowing of younger leaves; deformed fruit.	Buds die; flowers or fruit drops off.	Decrease in yield.
Manganese (Mn)	Interveinal chlorosis, white or gray 'spots' on younger leaves of small grains.	Stunted growth; leaves in vertical position.	Decrease in yield.
Copper (Cu)	Chlorosis of leaves; leaf tips turn white; appears on young leaves.	Poor growth, wilting.	Decrease in yield.
Calcium (Ca)	White strips along leaf margins; chlorosis on younger plant.	Death of buds and some roots; small leaves.	Decrease in yield.
Chlorine (Cl)	Chlorotic mottling of leaves.	Reduction in root growth.	Decrease in yield.
Molybdenum (Mo)	General pale green colour, and possible reddening of veins on the young leaves.	Stunted growth with small leaf size.	Decrease in yield.
Sulphur (S)	Yellow–green coloration in leaf blades and veins; appears on younger plant parts.	Slow growth; plants mature early.	Decrease in yield.

6 Characteristics of plant analysis

Plant analysis is based on the assumption that the nutrients in a plant are an indication of the supply of that nutrient. And, as such, plant nutrient levels are directly related to the amount available in the soil.

Since a shortage of a specific nutrient element will limit growth, other nutrient elements may accumulate in the cell sap. These will show at high levels in the test, regardless of soil supply.

Plant analysis

- Gives critical nutrient content determined more precisely.
- Adequate nutrient concentration is correlated to maximum yield

For the test to lead to effective soil health strategies being developed, a critical level has been determined for each element. A critical level is the content of an element below which the crop yield or performance will decrease below optimum.

There are two types of plant analysis

- **Tissue test** - made on fresh tissue in the field.
- **Total plant analysis** - performed in the laboratory with precise analytical procedures.

Drawbacks of using plant analysis to evaluate soil fertility.

- The test is affected by time of sampling and the plant part sampled.
- The deficiency will usually have caused yield loss by the time it can be detected.
- The crop may not yet have responded to the applied nutrient at the specific growth stage tested.
- Climatic condi-tions may be unfavorable for fertilization and/or for the crop to benefit from nu-trient additions.

7 Plant tissue analysis

Plant tissue analysis determines of nutrient content in fresh tissue in the field and implies the level of soil fertility. This test is rapid, but semi-quantitative.

Plant tissue analysis helps to

- Identify **nutrient deficiencies**
- Determine the **nutrient-supplying capacity** of the soil
- Determine the **effect of fertility treatments**

- show the **relationship between nutrient status** of plant and **crop performance.**

The most critical stage of sampling is the time of bloom, or from bloom to early fruiting time. It is important to understand that the level of other nutrients in the plant, and the soil or climatic conditions can affect the test results.

How to take sample for plant tissue analysis

Step 1 : Selecting sample

Sample most recently matured leaves

Note: nutrient concentration decreases with age so you must identify correct growth stage at sampling time.

Step 2 : Extracting sample

Extract the tissue from the sampled leaves or simply squeeze the leaves to produce sap.

Step 3 : Adding reagent

Add reagents to the sap to produce a coloured solution.

Step 4 : Interpreting the results

Compare the intensity of colour with standards on a colour chart. The test results may correlate with plant vigor and general performance.

Implications for extension The best time to do this test is at bloom. The limiting nutrient should be applied immediately. This will avert severe crop failure even though optimum yield will not be attained.

8 Total plant analysis

Total plant analysis is undertaken to determine total nutrient concentration in dried plant samples at a laboratory.

Precise analytical techniques are used for measuring the various elements from plant samples that have been specifically dried at regulated temperatures for the analysis.

The concentration of several elements can be measured simultaneously using the right equipment. Small difference in nutrient levels can be detected by this highly quantitative approach. Both assimilated and unassimilated nutrients are included.

Some characteristics of this approach are:

- Critical nutrient content is determined more precisely
- It is affected by time of sampling and the plant part
- Adequate nutrient concentration is correlated to maximum yield.

How to take sample for plant tissue analysis

- The part of the plant selected is of critical importance. Recently matured plant parts are the most reliable.

How to test tissue for nutrient deficiency

- Add reagents to the sap to produce a coloured solution.

How to interpret the results

The colour developed should be compared with standards on a colour chart that shows colours corresponding to specific nutrient deficiencies.

9 Other plant analysis approaches

a) Chlorophyll meters

Chlorophyll meters can be used to determine the plant nitrogen status.

The chlorophyll meters are:

- Quick
- Provide non-destructive analysis
- Give a good correlation with yield.

But, they must be calibrated for specific crops and growing conditions.

b) Grain analysis

Grain analysis is a post-mortem test as it is a test of the harvest grain.

- Good correlation with yield

10 Biological tests

a) **Field tests** consist of replicated small plots studies testing various treatments.

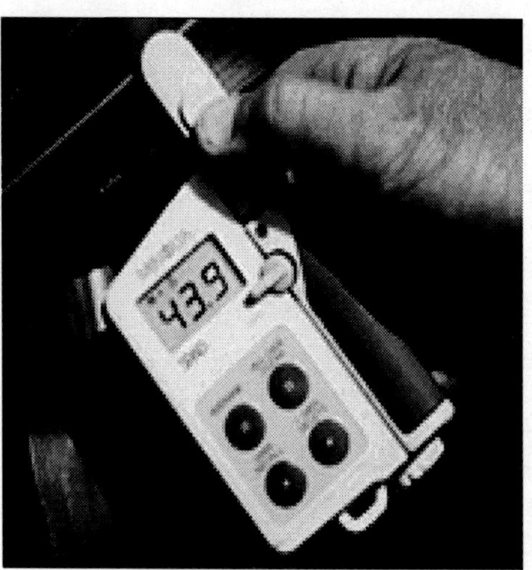

A chlorophyll meter

The results are indicative to similar soils with similar climatic conditions.

These experiments are helpful in formulating general recommendations. They are valuable tools but they are expensive, labour intensive and time consuming.

How to undertake a field test

Step 1 Layout the field in about 4-5 blocks (replicates) with a number of small plots equal to the number of treatments in each block. The blocks should be perpendicular to the soil fertility gradient (if visible) or slope.

Step 2 Allocate treatments to each plot by randomization

Step 3 Plant the whole trial in a day or at least a complete replicate in a day.

Step 4 Collect data on all plots and analyze statisticallyIt is important to follow approved methods for data collection and analysis

b) Laboratory and greenhouse tests

These are simpler and more rapid, using smaller volumes of soil in pots. There are several types.

How to undertake a laboratory or greenhouse test

Step 1 Soils samples are placed in pots and clearly labeled - these represent a wide range of chemical and physical properties

Step 2 Crop varieties are planted that are responsive to the specific nutrient being evaluated.

Step 3 Selected treatments are applied to the soil samples. For omission trials for NPK, omit one nutrient at a time such that you have treatments such as N0P0K0, N0PsKs, NsP0Ks, NsPsK0, NsPsKs, etc., where "s"= to sufficient amount of nutrient.

Step 4 Crop response to treatments is determined by measuring total plant yield and nutrient uptake.Predictions about increase in yield from a given amount of nutrient can then be derived.

11 Soil Testing

A soil test is a chemical method for estimating the nutrient supply ability of the soil.

Soil tests are done to

- Determine the relative ability of a soil to supply plant nutrients during a particular growing season.
- Predict the probability of obtaining a profitable response to fertilizer application.
- Determine the need to adjust soil pH.
- Diagnose problems of excessive salinity or alkalinity.
- Provide the basis for fertilizer recommendation for a particular crop.
- Evaluate the fertility status of the soil as the basis for planning a nutrient management programme.

A soil test result will give basic information about deficiencies and problems. These tests measure specific nutrients. These can be compared to the nutrients require for optimum yields and therefore suggest corrective actions that should be taken.

Soil tests are helpful diagnostic tool but do not provide absolute recommendations. The information they provide must be interpreted in relation to the goals and circumstances of the farmer.

A soil test only provides information about the fertility status of the soil. Correcting these is only one part of the farmer's crop management program. There are many factors that may result in low yields even when nutrients are adequate.

There are two types of soil test

- Laboratory analysis
- The use of mobile soil test kits

Taking a soil sample for laboratory testing

A soil test is only as good as the sample taken. The sample submitted for analysis must be representative of the field. Proper procedures must be followed when taking soil samples:

Step 1 Collecting the sample

- Take samples from the entire field
- Use soil auger or digging device and dig to specific depth (20 or 30 cm)

For a plot size of 1 hectare take about 20 samples and mix well together.

Step 2 Preparing the sample

One sample should be submitted for analysis if the field is considered uniform.

Taking the 20 sample (sometimes called the sub-samples):

- Remove all debris
- Drive the soil auger or sampling device to the specific depth and take sample
- Repeat process at all sampling points- take separate samples for portions of field that have different characteristics
- Mix all sub-samples thoroughly
- Take about 500 g for analysis and label with name, depth of sampling and date.
- Dry sample in shade or room on a brown paper.

Step 3 Submit one sample for analysis

The sample should be submitted as a dry sample in shade on paper (brown). Also re-bag dry soil in polythene bag for analysis

Some Laboratory Equipment for soil and plant analysis

12 pH testing

The importance of pH testing: Care must be taken to enhance and maintain soil structure and texture as well as ensure adequate levels of organic matter and balanced plant nutrition, i.e. crops need a variety of nutrients in a balanced proportion to ensure optimal growth. Also, managing such factors as soil pH and Cation Exchange Capacity (CEC) must be considered if nutrients are to be made available to plants.

Ovens should be used to dry plant samples at 65-70 °C

Digester and fume hood

Grinder

Vapodest distillation plant

Atomic Absorption Spectrophotometer

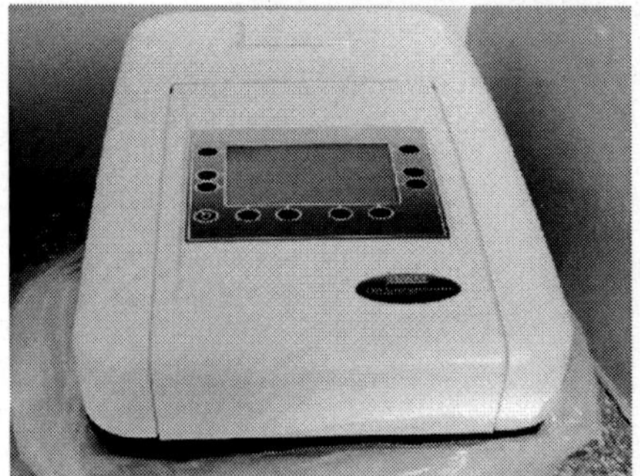

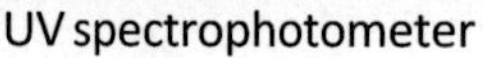

UV spectrophotometer

Flamephotometer

The pH of a soil controls many reactions in the soil so much so that the pH of a soil is also referred to as soil reaction.

As acidity increases or pH decreases

- The availability of basic cations like Ca^{2+} and Mg^{2+} decrease.
- AI becomes more toxic to plants causing root deformation.
- Availability of nutrients absorbed by plants in the anionic forms like phosphorus and molybdenum also decrease.
- Trace nutrient elements like Fe, Mn, Zn etc become more soluble and toxic to plants.
- Heavy metals like arsenic (As) and mercury (Hg) also become soluble and toxic.
- The activity of many soil organisms is reduced resulting in accumulation of organic matter, reduced decomposition and a lower availability of N, P and S.
- Nodulation of legumes decrease.

The **cation exchange capacity** is the ability of the soil to hold cations and exchange them with those in solution in a reversible chemical reaction.

Low pH is often considered a constraint to crop production. This is true where crops sensitive to low pH are grown (e.g. cotton). Low pH is associated with aluminium toxicity. Crops sensitive to aluminium toxicity are included in the cropping system or rotation.

Soil pH can be determined quite accurately in the field using a portable pH meter.

13 Yield gap analysis

The analysis of the yield reducing factors is referred to as yield gap analysis. When analyzing growth reducing and limiting factors, soil fertility is often one important factor.

Other factors include:

- Drought, sub-optimal water management practice or flooding
- Weed pressure, pests and diseases

A first approach to understand causes of low yields is to compare average yields in the village with the yields best farmers obtain.

Calculating the yield gap analysis 1

Table 7: Major effects of pH in the soil.

Factor	Effect
Al toxicity	Al toxicity decreases with increasing pH.
P availability	P availability is greatest from pH 5.5-7.D
Micronutrient availability (nutrients required in small amounts by plants)	All micronutrients, except Mo, are more available from pH 5.5-6,0 (Mn and Fe toxicity is minimized in this range).
Cation exchange capacity (the ability of a soil to retain cations such as Ca, Mg, K)	The cation exchange capacaity increases with increasing pH in highly weathered soils. This means the soil is able to retain more Ca, Mg, K, which might otherwise be lost to leaching.
Nitrogen mineralization (the release of N from organic into plant-available forms) N_2-fixatiorr (the conversion of N from the atmosphere into forms that can be used by plants)	Soil organisms required for Nmineralization function best at soil pH 5,5-6.5. Infixing nodules are less likely to occur and function less effectively at pH > 5.0.
Disease	Some diseases can be controlled by manipulating soil pH (e.g., potato scab incidence decreases with decreasing pH).
Rock phosphate (RP) dissolution	Soil pH must be < 5,5 for RP to dissolve and release P for plant uptake.

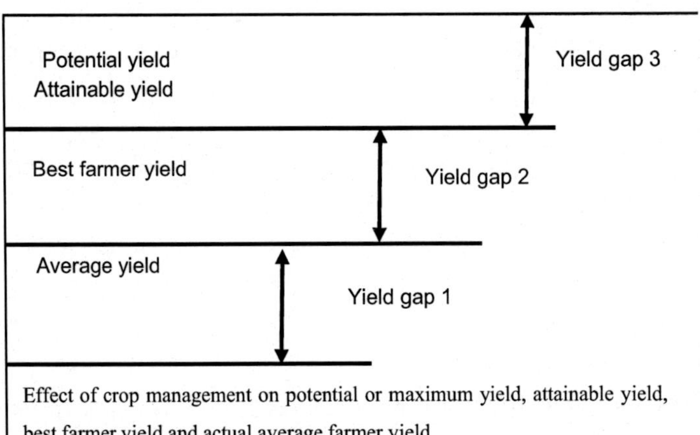

Effect of crop management on potential or maximum yield, attainable yield, best farmer yield and actual average farmer yield

Step 1 Find out what "best farmers" do differently. This will help identify the causes of the differences, e.g. weed, pests or diseases, level of fertilizer application (reducing factors), and will also provide the scope for short-term improvement

Step 2 Calculate the average yield

Step 3 Calculate the yield gap 1 Best farmer yield - average yield = yield gap 1).

a) The genetic yield potential

For a given crop cultivar in a particular climate, there is a potential yield that is determined solely by carbon dioxide (CO_2), solar radiation and crop characteristics.

The genetic yield potential is the maximum yield that can be obtained based on the assumption that only two factors have an impact - the climate and crop cultivar. Water and nutrients are considered not limiting.

Farmers often obtain less than 50% of this genetic yield potential for a given planting date, cultivar choice and site.

b) The nutrient-limited yield

The **attainable, or nutrient-limited, yield** is what farmers can achieve with typical soil fertility management practices but with optimal water and crop management.

In reality, farmers obtain much less than the attainable yields due to a range of constraints to crop growth.

Low farm level yields compared with potential Soil fertility is one of the growth-reduction and limiting factors. Other factors include too much or too little water, pests, diseases, etc. See figure.... below. Current management practices may prevent the farmer from obtaining better yields. Smallholders may **not** have:

- The right crop
- The right crop variety
- The right plant population
- The right level of organic matter in the soil
- Sown the seed at the right date
- Applied the right type and right quantity of fertilizer.

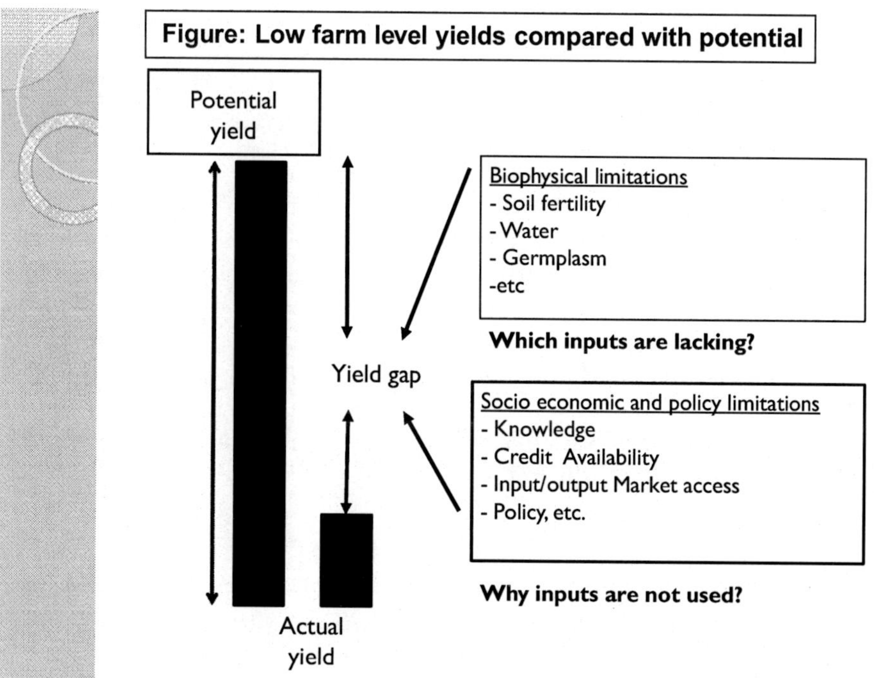

c) Nutrient flow analysis

Nutrient flow analysis can be used to determine the impact of farmer management decisions on soil fertility on the farm.

Farmers remove harvested products, straw or crop residue for building or handicrafts, residues or wood for fuel, from their smallholdings. These processes lead to a loss in nutrients, e.g. burning of straw will lead to complete loss of carbon (C) and nitrogen (N).

Estimating nutrient flows is a useful way to find out if farmers' crop management practices are sustainable. Put another way, are outputs of nutrients balanced by a sufficient level of inputs of organic and inorganic fertilizer or from growing legumes.

There is need to express nutrient amounts in the same unit to be able to compare flows, e.g. kilograms of nitrogen (N), phosphors (P) and potassium (K).

This requires that the concentration of N, P and K in all plant parts, manure, fertilizer, are determined.

Nutrient flow analysis should enable a farmer answer questions such as "what is happening to my soil if I do not apply fertilizer to my maize field, and I sell the grain and use the stover for fuel wood"?

Implications for extension

Extension worker can undertake yield gap analysis without needing many resources.

Studying what the most productive farmers are able to achieve and what they are doing can be a great way to kick start agricultural productivity. It will also mean that there are demonstration sites and champions for different ways of working. Sometime genuine farmer-led innovations are identified often from outside social or cultural norms - for example the farmer that doesn't engage in burning fields; or the farmers that don't clear away weeds or cut legumes at the root rather than pulling them out. These may be seen as wrong - but actually build up the organic matter in the soil.

Other Methods

14 Resource flow mapping

Resource flow mapping helped farmers in learning to

- Properly plan for their limited land resources e.g. through crop rotation.
- Make priorities and allocate resources for fertility improvement.

- Allocate labour resources to various farm enterprises and in a timely manner.
- Estimate quantities of resources entering (inputs) and leaving (outputs) the farm every season.
- Reflect on balancing their resource use against their resource needs.

The map shows three main destinations for nutrients

- Crop produce sold at market, i.e. nutrients definitely removed from the farm.
- Domestically consumed crop produce - nutrients ending up in pit latrines and household refuse.
- Crop residues removed from the field in various forms and taken to the compound to be used for other purposes, e.g. fuel, animal fodder, thatching and composting material. These maps then formed a basis for nutrient flow analysis using the resource kit computer package.

15 Computer-based diagnostic tools

Computer-based diagnostic tools such as the use of computer simulation models including:

a) NUTMON Low cost software - but highly demanding in terms of information needed for the model

NUTMON is described as a toolbox was developed model at the Wageningen University and Research Center in close collaboration with institutions in Kenya, Uganda and Burkina Faso.

NUTMON is an integrated, multidisciplinary methodology, technique where farmers and researchers jointly analyse the environmental and financial sustainability of tropical farming systems.

The NUTMON approach is being implemented in research and development projects addressing soil fertility management in situations of nutrient depletion or nutrient accumulation in Ghana.The NUTMON-Toolbox consists of a questionnaire, a manual and several software modules that are specifically designed to facilitate monitoring and analysis of nutrient flows and economic performance at farm level.

Participatory research techniques are used to obtain the farmers perspective including:

- Resource flow mapping
- Matrix ranking
- Trend analysis.

The software also helps to generate also quantitative analysis tools that generate important indicators including:

- Nutrient flows
- Nutrient balances
- Cash flows, gross margins and farm income.

Both the qualitative and quantitative analysis are then used to improve or design new technologies which tackle soil fertility management problems and which can help to increase the financial performance of the farm.

NUTMON requires a substantial amount of data:

Soil: Carbon (C), nitrogen (N), phosphorous (P) and potassium (K) contents, bulk density, slope, mineralisation rate, rootable depth, enrichment factor and erodibility

- **Weather**: monthly rainfall, rainfall erosivity (USLE R-factor)
- **Crop**: type, area, yield (grain, straw), destination of products, crop calendar
- **Animals:** type, growth and composition, production, livestock confinement per month
- **Redistribution units**: size and quality of latrines, compost pits, manure heaps etc.
- **Management:** internal and external inputs per field, animal and redistribution units. In addition, information is required about nutrient contents of all products, prices, feed requirement, production of human and animal excreta, production of household.

b) **NUTRIENT MANAGEMENT SUPPORT SYSTEM (NuMaSS)**

NuMaSS is a tool that diagnoses soil constraints and selects appropriate management practices, based on agronomic, economic and environmental criteria for location specific conditions.

NuMaSS integrates three existing nutrient decision support systems:

- Acid Decision Support System (ADSS)
- Nitrogen Decision Support System (NDSS) and
- Phosphorus Decision Support System (PDSS).

The three systems are integrated as modules into one system with a shared interface. NuMaSS consists of five sections:

- Geography: distinguishes between humid / tropical, wet / dry and semi-arid
- Diagnosis: provides an early indication whether there is a nutrient management problem diagnoses soil constraints:
  - Intended crop (name, target yield and Al-saturation must be provided)
  - Previous cropping
  - Soil order (USDA taxonomy). Default values for soil data is available
  - Plant: plant analysis, nutrient deficiency symptoms, indicator plants.

  The result of the diagnosis is the likelihood of an acidity, nitrogen or phosphorus constraints.
- Prediction: organic application, lime application, nutrient application.
- Economics
- Results

c) **QUEFTS (**Quantitative Evaluation of the Fertility of Tropical Soils)

QUEFTS was developed in the late 1980's. The developers claim QUEFTS can contribute to a more efficient procurement and use of mineral fertilizers at both regional and farm level. It is a tool for quantitative evaluation of the native fertility of tropical soils, using calculated yields of unfertilized maize as a yardstick. The QUEFTS model works only for a maize crop. The generic version of QUEFTS is called CROPFERT and can be used for all type of crops, for which the nutrient concentrations are given in the file NUTRIDAT.dat

QUEFTS is a four-step process

Step 1 chemical soil test values,

The potential supplies of nitrogen, phosphorus and potassium are calculated, applying relationships between chemical properties of the

0-20 cm soil layer and the maximum quantity of those nutrients that can be taken up by maize, if no other nutrients and no other growth factors are yield-limiting.

Step 2 potential NPK supply from soils and fertilizer

The actual uptake of each nutrient is calculated as a function of the potential supply of that nutrient, taking into account the potential supplies of the other two nutrients.

Step 3 actual NPK uptake

Comprises the establishment of three yield ranges, as depending on the actual uptakes of nitrogen, phosphorus, and potassium, respectively.

Step 4 maize grain yield - acknowledging interactions between the three macronutrients

The yield ranges are combined in pairs, and the yields estimated for pairs of nutrients are averaged to obtain an ultimate yield estimate.

d) **DSSAT**

DSSAT (The Decision Support System for Agrotechnology Transfer) is a software application that comprises crop simulation models for over 28 crops.These include:

- Bare fallow
- Bell (pepper, cabbage, potato & tomato
- Chickpea, cowpea, dry bean, faba bean, peanut, soybean & velvet bean
- Barley, maize, millet, rice, sorghum & wheat.

It simulates growth, development and yield as a function of the soil-plant-atmosphere dynamics. It can be used for on-farm and precision management to regional assessments of the impact of climate variability and climate change. It has been in use for more than 20 years in over 100 countries.

The crop models require daily weather data, soil surface and profile information, and detailed crop management as input. Crop genetic information is defined in a crop species file that is provided but cultivar, or variety, information, is provided by the user. Simulations are initiated either at planting or prior to planting through the simulation of a bare fallow period. These simulations are conducted at a daily step and, in some cases, at an hourly time step depending on the process and the crop model. At the end of the day the plant and soil water, nitrogen and carbon balances are updated, as well as the crop's vegetative and reproductive development stage.

DSSAT integrates the effects of soil, crop phenotype, weather and management options, and allows users to ask "what if" questions by conducting virtual simulation experiments on a desktop computer in minutes which would consume a significant part of an agronomist's career if conducted as real experiments.

DSSAT also provides for evaluation of crop model outputs with experimental data, thus allowing users to compare simulated outcomes with observed results. This is critical prior to any application of a crop model, especially if real-world decisions or recommendations are based on modelled results. Crop model evaluation is accomplished by inputting the user's minimum data, running the model, and comparing outputs with observed data. By simulating probable outcomes of crop management strategies, DSSAT offers users information with which to rapidly appraise new crops, products, and practices for adoption.

DSSAT has been developed through collaboration between scientists at the University of Florida, the University of Georgia, University of Guelph, University of Hawaii, the International Center for Soil Fertility and Agricultural Development, Iowa State University and other scientists associated with the International Consortium for Agricultural Systems Applications (ICASA).

e) APSIM

The Agricultural Production Systems Simulator APSIM

APSIM is an advanced simulator of agricultural systems using modified open source software. It contains a suite of modules that enable the simulation of systems that cover a range of plant, animal, soil, climate and management interactions.

Crop coverage includes

- Various pasture crops.
- Barley, maize, millet, rice (ORYZA), sorghum & wheat.
- Chickpea, cowpea, faba bean, fieldpea, lablab, mucuna, mungbe navybean, peanut, pigeonpea & soybean.

Chapter 3

Water Management

Summary

This chapter explores different approaches to manage water in rain-fed and irrigated agricultural systems. The first methods support the retention of water and reduce soil erosion. These methods include mulching – the application of a surface coating over the soil – which is usually organic mater. This reduces the damage to soil caused by rain and wind and reduces water loss to evaporation and helps with infiltration. The other methods explored are bunding and terracing. These approaches are about stopping the flow of water across land to increase the amount of water infiltration into the soil and to reduce or prevent soil run off. These approaches are labour intensive. A less labour -intensive approach is contour ploughing. This approach is about encouraging farmers and tractor drivers to plough across sloping fields not up and down them. The second group of technologies refers to the introduction of water through irrigation methods.

1 Mulching

Mulching refers to covering the ground with organic material such as crop residues, or with other materials such as plastic. Mulching can either be live or dead mulch.

For cover crops see Chapter 4 section 11.

Surface mulching: Surface mulching is a useful field practice in terms of soil surface protection and water use efficiency, but one that is difficult to achieve. Crop residues, suitable for mulching, have competing uses and are subject to rapid loss by termites and other soil fauna. Surface mulches can be subjected to rapid removal and comminution loss - and this reduces their intended purpose. Mulch can also be obtained from pruning cut from boundary areas and natural vegetation but this operation is labour consuming and the prunings are often better used on higher value crops, for animal feeds or as material for compost making. Near permanent soil cover is one of the foundations of conservation agriculture.

(see Chapter 4 section 1b). The simple irrigation approaches channel surface or flood water to crop fields where the water can be used.

Comminution is the process in which solid materials are reduced in size, by crushing, grinding and other processes.

The importance of mulching in soil fertility management includes:

- Protecting the soil particles from being carried away by strong winds, thereby decreasing wind erosion.
- Protecting the soil from forming a crust.

Structural soil crusts are relatively thin, dense, somewhat continuous layers of non-aggregated soil particles on the surface of tilled and exposed soils. A surface crust is much more compact, hard and brittle when dry than the soil immediately beneath it, which may be loose and friable. See figure below.

(Soil crust: courtesy google images)

A surface crust indicates poor infiltration, a problematical seedbed, and reduced air exchange between the soil and atmosphere. It can also indicate that a soil has a high sodium content that increases soil dispersion when it is wetted by rainfall or irrigation.

- Reduces the direct impact of raindrops on the bare soil (the kinetic energy of the rain), because it allows rainwater to infiltrate and thus minimize soil erosion from water.
- Protecting the soil from becoming dehydrated. Together with increased infiltration, this ensures that the moisture content in the soil remains higher than in soil without a mulch layer. It will thus take longer in the dry season for crops with a mulch layer to be short of water.
- It suppresses weed growth – and therefore competition for nutrients and water

- When organic matter is used as mulch it will slowly breakdown releasing organic matter into the soil.

Bare Soil

Mulched Soil

It is important to ensure that mulches are not capable of broadcasting seeds or create weed growth on the plot.

2 Bunding

A bund is an embankment or heap of earth or stone. It is a simple method in soil and water conservation employed in a field with medium slope.

The bunds are built using stone or earthen walls across a slope in order to serve as barrier to runoff.

Building soil conservation bunds

When they are built along the contours they are referred to as contour bunds. In order to strengthen the bunds and increase its stability trees, which fix nitrogen, are planted along bunds in addition to grass.

Bunds help to improve soil fertility in the following ways:

- As water flow is obstructed by bunds, rate of water infiltration is increased.
- Soils along the sloppy fields are protected from erosion caused by flowing water.
- Moisture is retained in the soil hence increasing the water holding capacity of the soil.
- Biomass is generated along the bunds which can be composted and used as organic matter.

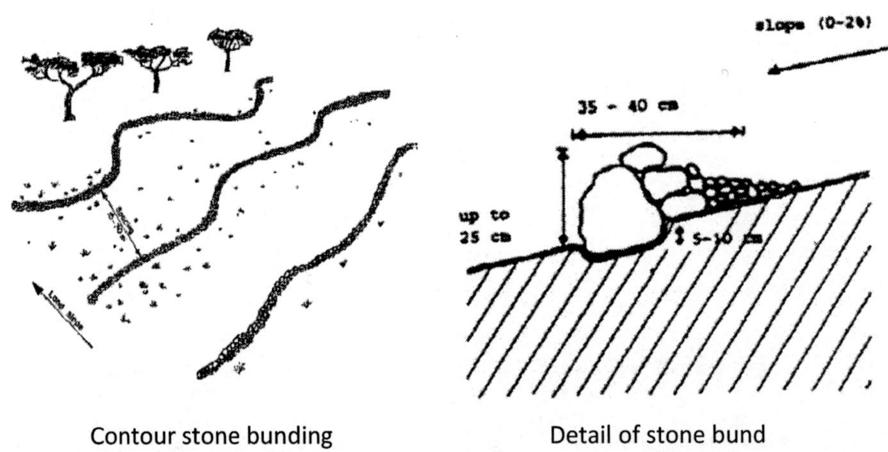

Contour stone bunding   Detail of stone bund

3 Terracing

A terrace is a piece of sloped land that has been cut into a series of successively receding flat surfaces or platforms, which resemble steps, for the purposes of more effective farming. This type of landscaping, therefore, is called terracing.

On sloping lands, terracing is necessary for reducing overland flow rates thereby, reducing erosion to conserve nutrient on the soil surface. Terraces also help to retain water thereby increasing the water holding capacity of the soil.

The faced structure is referred to as a riser, bank, dyke, ridge, wall or embankment.

A terraced field

There are two types of terraces:

- **Graded terraces** are used in areas with abundant rainfall to divert excess surface water.
- **Level terraces** are used in areas with permeable soils and low to moderate rainfall

4 Contour ploughing

Contour ploughing is an effective water management option for ISFM in semi-arid zones.

It is the farming practice of ploughing across a slope following its elevation contour lines. The rows form slow water run-off during rainstorms to prevent soil erosion and allow the water time to settle into the soil. In contour ploughing, the ruts made by the plough run across the slopes, generally resulting in furrows that curve around the land and are level.

Land ploughed along the contours

Implications for extension: The application of mulch is a relatively low cost approach which can reduce labour needs for weeding. However there is competition for the organic materials that can be used as mulch – and other materials may prove too expensive.

Contour ploughing involves encouraging tractor drivers to plough across a slope not up and down it. This means that water will be retained longer along the plough ridges. The creation of bunds, or earth mounds, uses the same principle but is more labour intensive – especially to introduce them where they haven't existed before.

Extension officers need to able to understand the difference between growing rice in a paddy and the role bunds can play in retaining some water – but not standing water. Rice is not affected by standing water but other cereal crops are such as maize are very vulnerable to waterlogging.

The creation of terraces is a huge commitment of time and resources that is likely to be beyond most smallholder's resources. Examples of terracing can be found in the Nangodi area in the Upper East Region of Ghana.

5 Irrigation

Irrigation may be defined as the science of artificial application of water to the land or soil. It is used to assist in the growing of agricultural crops, maintenance of landscapes, and re-vegetation of disturbed soils in dry areas during periods of inadequate rainfall.

A portion of land can be irrigated during the dry season when there is less than average amounts of rainfall or when the land does not receive enough water on its own to be fertile.

Water that is used for irrigation may be taken from nearby lakes, reservoirs, rivers or wells. The amount of water that is to be used for irrigation depends on the type of crop that is being farmed as well as the amount of rainfall received in the region where the farm is located.

6 Surface irrigation (flood irrigation)

In surface irrigation systems, water moves over and across the land by simple gravity flow in order to wet it and to infiltrate into the soil. It is often called flood irrigation when the irrigation results in flooding or near flooding of the cultivated land. Historically, this has been the most common method of irrigating agricultural land.

Surface irrigation methods

Furrow irrigation is conducted by creating small parallel channels along the field length in the direction of predominant slope. Water is applied to the top end of each furrow and flows down the field under the influence of gravity. Water may be supplied using gated pipe, siphon and head ditch or bankless systems. The speed of water movement is determined by many factors such as slope, surface roughness and furrow shape but most importantly by the inflow rate and soil infiltration rate. The spacing between adjacent furrows is governed by the crop species, common spacings typically range from 0.75-2 metres. The crop is planted on the ridge between furrows that may contain a single row of plants or several rows in the case of a bed type system. Furrow irrigation is particularly suited to broad-acre row crops such as cotton, maize and sugar cane. It is also practiced in various horticultural industries such as citrus, stone fruit and tomatoes.

Basin irrigation: level basin irrigation has historically been used in small areas having level surfaces that are surrounded by earth banks. The water is applied rapidly to the entire basin and is allowed to infiltrate. Basins may be linked sequentially so that drainage from one basin is diverted into the next once the desired soil water deficit is satisfied. A "closed" type basin is one where no water is drained from the basin. Basin irrigation is favoured in soils with relatively low infiltration rates. Fields are typically set up to follow the natural contours of the land but the introduction of laser levelling and land grading has permitted the construction of large rectangular basins. Basin irrigation is commonly used in the production of crops such as rice and wheat.

Border strip or bay irrigation could be considered as a hybrid of level basin and furrow irrigation. The borders of the irrigated strip are longer and the strips are narrower than for basin irrigation and are orientated to align lengthwise with the slope of the field. The water is applied to the top end of the bay, which is usually constructed to facilitate free-flowing conditions at the downstream end. One common use of this technique includes the irrigation of pasture for dairy production.

7 Localized irrigation

Localized irrigation is a system where water is distributed under low pressure through a piped network, in a pre-determined pattern, and applied as a small discharge to each plant or adjacent to it.

Drip irrigation releases water in drops just at the position of roots. Water is delivered at or near the root zone of plants, drop by drop. This method can be the most water-efficient method of irrigation, if managed properly because it minimizes evaporation and runoff. Drip irrigation systems are often designed to operate daily for nearly the entire day and to supply water to only the root zone of the crop.

Sprinkler irrigation uses pressure energy to form and distribute "rain like droplets" over the land surface. Although they are normally designed to supply the irrigation requirements of the farm, sprinkler systems are also used for crop and soil cooling, frost protection, delaying fruit and bud development, application of agricultural chemicals, and land application of waste waters. In a sprinkler system, water is conveyed from a pump or other source of water under pressure through a network of pipes, called main lines and sub-mains, to one or more pipes with sprinkler called laterals. The sprinkler distributes the water over the land surface.

Sub-irrigation, also sometimes called seepage irrigation, has been used for many years in field crops in areas with high water tables. It is a method of artificially raising the water table to allow the soil to be moistened from below the plant's root zone. These systems are often located on permanent grasslands in lowlands or river valleys and combined with drainage infrastructure. A system of pumping stations, canals, weirs and gates allows it to increase or decrease the water level in a network of ditches and thereby control the water table.

Lateral move (side roll, wheel line) A series of pipes, each with a wheel of about 1.5m diameter permanently affixed to its midpoint and sprinklers along its length, are coupled together at one edge of a field. Water is supplied at one end using a large hose. After sufficient water has been applied, the hose is removed and the remaining assembly rotated either by hand or with a purpose-built mechanism, so that the sprinklers move 10m across the field. The hose is reconnected.

The process is repeated until the opposite edge of the field is reached. This system is less expensive to install but much more labour intensive to operate, and it is limited in the amount of water it can carry. One feature of a lateral move system is that it consists of sections that can be easily disconnected. They are most often used for small or oddly shaped fields, such as those found in hilly or mountainous regions, or in regions where labour is inexpensive.

Implications for extension

Irrigation is one of the most expensive interventions in agriculture. It is not an intervention that smallholders can usually consider. As an extension officer you may come across existing schemes or be involved in government or development agency investment in irrigation so it is important to understand the different approaches.

Surface, or flood irrigation, systems come in three different types; furrow, basin and border strip. In addition there are a series of localized irrigation approaches; drip and sprinkler systems, sub-irrigation and lateral move systems. Each has different merits for different applications.

Chapter 4
Approaches to Soil Fertility Management

Summary

This chapter looks at the various approaches available to the farmer for the management of soil fertility. In general there are four approaches, namely, physical, biological and chemical approaches as well as the cropping systems or agronomic approach.This chapter discusses three approaches, physical, biological and cropping system or agronomic approaches. Chemical approach has been treated in chapter 7 of this book. Inoculation is an important technique in legume production. This chapter also looks at inoculation and how it is done.

1 Physical approaches to soil fertility management

Physical processes are the mechanical and/or manual practices that improve soil fertility. There are many different - but they include tillage, mulching, bunding and terracing. These last three are covered in section 3.

❒ Conventional tillage

Tillage is defined as any physical loosening of the soil carried out in a range of cultivation operations, either by hand or mechanized. This can include any process that breaks the soil – this can include hoeing, or harvesting.

Tilling or ploughing is used to remove weeds, shape the soil into rows of crop plants and furrows for irrigation. This leads to unfavorable effects, like soil compaction, loss of organic matter, degradation of soil aggregates, death or disruption of soil microbes and other organisms including mycorrhiza, arthropods and earthworms. Soil erosion, where topsoil is blown or washed away is also a negative consequence of tillage.

Where tillage is done effectively it can break up hard pans and allow crop routes to access nutrients that otherwise would not be available. Bad practices in tillage can see topsoil buried below the depth of plant roots - making the nutrients inaccessible.

Conventional tillage hand hoeing or ploughing breaks the soil – the way this is practiced in Ghana this approach leaves only 15% or less crop residue on the field.

- Ploughing loosens and aerates the top layer of soil that can facilitate the planting of crops.
- Tillage improves soil structure by reducing soil compaction.

Conservation tillage or conservation agriculture

Conservation agriculture (sometimes called conservation tillage or zero tillage) is underpinned by three basic principles:

- It is a way of growing crops from year to year with minimized soil disturbance - seeds are drilled into soil.
- **The soil is kept covered** with organic materials (crop harvest residues or cover crops) – at least 30% soil cover (zero tillage aims at 100% ground cover) – mulching reduces water evaporation and erosion
- **Crop rotations**/associations are used.

All three principles are important possible options for smallholders, as with all techniques, their use should depend on a case-by-case assessment of local requirements as part of local adaptation. Full conservation agriculture requires a fundamental change in farming systems that may not be practical or economic for the farmer.

Advantages of conservation agriculture

- Rapid planting of large areas.
- Zero tillage provides ideal environment that encourage the growth of some microorganism and by their burrowing activities, other soil organisms create pores that help in aeration and infiltration. It increases the amount and variety of life in and on the soil.
- Reduction or elimination of soil erosion. The practices often leave substantial amounts of crop residue evenly distributed over the soil surface to protect the soil against the potential of rainfall's kinetic energy to generate sediment and decrease water runoff.

Pitfalls that have been encountered

- When conservation agriculture implementation has been attempted without sufficient local adaptation:

- Insufficient crop residues available for mulching where they provide greater returns for the farmer when used as animal feed.
- Short-term decreases in crop yields may be experienced as significant crop yield increases are only achieved in the long term.
- May increase labour requirements unless herbicides/ weedicides are available and cost effective for weed control.
- Conservation agriculture may not result in improved agronomic efficiency of fertilizer use.
- Agronomic efficiency may even be reduced under prolonged conservation agriculture due to increased leaching because of increased water infiltration and a more continuous macropore system in the soil.
- Zero tillage requires different equipment and significant changes to inputs and agronomic practices.

One of the claims for conservation agriculture is the potential to build up soil organic carbon. This depends on soil texture (particularly clay content) and the extent to which the soil's capacity to store carbon has already been reached.

Surface ***runoff*** is the water flow that occurs when the soil is infiltrated to full capacity and excess water from rain, snow melt or other sources flows over the land. It occurs when there is more water than land can absorb.

Implications for extension Land preparation and tillage can be a huge burden to agricultural households.

The brunt of hand hoeing tends to be undertaken by women. Mechanization changes this – as most tractor drivers tend to be men – however, the burden on women can increase with mechanized ploughing because often a larger area is prepared and this means more weeding, harvesting and post-harvest treatments.

Extension workers can add value by helping to bundle farms for mechanized tillage – so that fuel is not wasted driving between smallholdings – but rather adjoining plots are ready and ploughed or tilled at the same time.

Like all agricultural innovations – conservation agriculture requires local adaptation.

2 Biological approaches to soil fertility

Biological approaches deal with the manipulation of biological processes. This approach includes enhancing the potential of legume crops to fix nitrogen into the soil.

What is biological nitrogen fixation (BNF)?

BNF is the process that changes atmospheric nitrogen, which is inert, to a biologically useful form. Chemically, this is similar to the Haber-Bosch process, which is the principal commercial method of producing ammonia for nitrogenous fertilizers, by direct combination of nitrogen and hydrogen under high temperature and pressure.

The BNF process is mediated much more efficiently in nature by bacteria. Plants benefit from this process when the bacteria die and release the nitrogen to the soil or when the bacteria live in very close association with plants.

Bacterial associations with certain plant families, primarily legume species, make the largest single contribution to biological nitrogen fixation in the biosphere.

In leguminous crops (e.g. common bean or soybean), the bacteria, collectively called rhizobia, live in nodules where the plants fix the nitrogen to ammonia, which is absorbed by the plants. Thus, there is a close symbiotic relationship between the plants and the bacteria. While there is a wide range of organisms and microbial-plant associations that are capable of fixing atmospheric N2, the symbiotic relationship between rhizobia and legumes is responsible for contributing the largest amounts of fixed nitrogen to agriculture.

Grain legumes contribute more than 20 million tons of fixed N to agriculture each year (Herridge et al. 2008). The impact can be dramatic at smallholder farmer level. In 2010 a programme called N2 Africa estimated, using inoculants, it would raise average grain yields by 954kg/ hectare in four legumes (groundnut, cowpea, soybean, and common bean), increase average biological nitrogen fixation by 46kg/hectare and increase average household income by US$465, in its target countries. So an important capacity building step for smallholder farmers in Africa is to select superior rhizobia strains for enhanced biological nitrogen fixation (BNF) and develop inoculum production capacity in sub-Saharan Africa, with public and private sector partners.

Thus, the significance of BNF as the major mechanism for the recycling of nitrogen from the atmosphere to available forms in the biosphere cannot be overemphasized.

The various biological approaches are presented as:

Green manuring

Green manuring consists of ploughing in green, non woody plants or plant parts. The plant material can come from a crop that was grown after or between the main crops, or from a weed that grew during a fallow period. It can also come from a shade plant or tree whose cuttings or fallen leaves are suitable for ploughing into the soil. Adding green manure to the soil has several advantages.

- Green manuring make nutrients available for the main crop.
- Green manure improves the soil structure.
- It increases or retains the level of organic matter in the soil.
- Green manure increases the ability of the soil to retain moisture.
- Protect the soil against rain and wind erosion, dehydration and extreme temperature fluctuations at a time when no other crops are present.
- When using leguminous plants as green manure they fix extra nitrogen out of the air, which becomes available to the main crop after the manure has been ploughed into the soil.

Cover cropping

These are crops grown to cover the area between rows of commercial crops or to cover the whole field when no commercial crops are growing. Eg. *Cajanus cajan.* The contributions of cover crops to soil fertility are:

- Cover crop reduces the kinetic energy of the rain drops and thereby reduces the erosive potential of rainfall.
- Most cover crops especially the leguminous ones are rich in Nitrogen which is released to the soil gradually. This reduces the rate of leaching of Nitrogen and ensures prolonged availability of the nutrient in the soil.
- Some cover crops have powerful tap roots which can break up soil compaction thus de-compacting soils.
- Cover crops adds organic matter to the soil when they die thereby improving the structure, infiltration and water holding capacity of the soil.

Manuring

Manure consists of animal excrement, usually mixed with straw or leaves. The amount and quality of the excrement depend on the animals feed. Good manure contains more than just excrement and urine. Straw and leaves are added. Using aged manure is an ideal method to retain and increase soil fertility. The following are some of the importance of manures in soil fertility management.

- Manure increases the level of organic matter in the soil.
- Manure increases the available nutrients in the soil.
- Manure stimulates soil organisms which improves the soil structure (aggregate formation) and water retention capacity of the soil.

Incorporation of crop residue management

Crop residue management is a conservation practice that usually involves a reduction in the number of passes over the field with tillage implements and/or in the intensity of tillage operations, including the elimination of ploughing (inversion of the surface layer of soil). This practice is designed to leave sufficient residue on the soil surface to reduce wind and/or water erosion. CRM is a year-round system that includes all field operations that affect the amount of residue, its orientation to the soil surface and prevailing wind and rainfall patterns, and the evenness of residue distribution throughout the period requiring protection. This may include the use of cover crops where sufficient quantities of other residue are not available to reduce the vulnerability of the soil to erosion during critical periods.

Having the crop residue on the field is not all that matters but proper management of the residue is the key to attaining the maximum benefits. Managing the residue entails spreading the residue uniformly on the entire soil surface. The residue contributes to soil fertility in the following ways.

- The incorporation of crop residue adds organic matter to the soil which improves soil structure and water infiltration.
- Crop residue provides ideal environment that encourage the growth of some micro organism. Also, by their burrowing activities, soil organisms create pores that help in aeration and infiltration.
- Crop residues retain Carbon (C) in the soil.
- It buffers soil pH and facilitates availability of nutrients.
- Crop residues reduce evaporation and increases soil moisture.

Zero tillage

Zero tillage or no tillage is a way of growing crops from year to year without disturbing the soil through tillage. Zero tillage aims at 100% ground cover. Residues, weeds, equipments, crop rotations, water, diseases, pests and fertilizer management are just some of the many details of farming that change when switching to zero tillage. Tilling is used to remove weeds, shape the soil into rows of crop plants and furrows for irrigation. This leads to unfavorable effects, like soil compaction, loss of organic matter, degradation of soil aggregates, death or disruption of soil microbes and other organisms including mycorrhiza, arthropods and earthworms. Soil erosion, where top soil is blown or washed away is also a negative consequence of tillage.

- Zero tillage is an agricultural technique which increases the amount of water and organic matter (nutrients) in the soil and decreases erosion. It increases the amount and variety of life in and on the soil but may require herbicide usage.
- Zero tillage provides ideal environment that encourage the growth of some micro organism and by their burrowing activities, other soil organisms create pores that help in aeration and infiltration.
- Zero tillage can increase yield because of higher water infiltration and storage capacity and less erosion.
- Zero tillage improves soil quality, aggregation and reduces water evaporation.
- Zero tillage has carbon sequestration potential through storage of soil organic matter in the soil of crop fields.

Non-burning

This is an act of not using fire as a management tool on a piece of land. This act helps to protect the soil in many ways and does not leave the soil bare. It also ensures the following:

- Non-burning, erosion is reduced and hence, nutrients are available in the top soil.
- Microorganisms are not destroyed hence there is increased in infiltration through their burrowing activities.
- The presence of vegetation results in conservation of water and reduced evaporation rates.

- Crop residues and grasses can be worked into the soil to increase the organic matter content of the soil. This will in tend improve soil structure.

3 Use of inoculants

Inoculation of grain legumes with rhizobia is an important process to maximize biological nitrogen fixation capacity in these crops. Inoculation enhances the ability of leguminous crops to fix nitrogen. It has the potential of increasing dry matter yield (the dry weight of material produced), and residual nitrogen levels.

Nodulating bacteria such as rhizobia are introduced to legume seeds before planting. Inoculation of grain legumes with rhizobia is an important process to maximize biological nitrogenfixation capacity in these crops. Inoculation has the potential of increasing dry matter yield, nitrogen yield, and residual nitrogen levels

Un-inoculated Soybean   Inoculated Soybean

Illustration of nodule formation in inoculated and un-inoculated soybean

The benefits of using an inoculant in legumes include

- Higher crop yield
- Increased protein content of the seed
- Availability of more soil nitrogen for other crops.

Inoculant storage, handling and application

The carrier medium should protect the rhizobia in the package and on the seed. Inoculant should be packaged to protect the rhizobia until it is used. The package should allow exchange of gases and retention of moisture.The package should provide clear instructions and list the legumes that it effectively nodulates and carry an expiry date beyond which the product cannot be considered dependable.

Legume inoculants are perishable and quickly lose their effectiveness when exposed to a temperature of 40°C or more. Inoculants retain their effectiveness for six months or longer when stored at a temperature around 20°C. This period can be extended if refrigerated near 4°C but freezing inoculants damages the product.

Simple precautions prior to inoculant application and planting protects rhizobia

Inoculants should

- **Never** get too hot (over 30°c)
- **Never** dry out
- **Not** be used after their expiry date (However, conservative expiration dates will have been set to protect the interests of users).

Smallholders should

- **Never** mix fertilizer with inoculated seed (no direct contact)
- **Never** broadcast inoculants onto dry soil
- **Never** apply additional inoculant to the surface when the soil is dry

Inoculation process

1. Weigh 1g (2 Voltic water bottle top) of inoculum. DO NOT PRESS OR FIRM.

2. Weigh 1 kg of seed
3. Sprinkle water on seed to make it moist
4. Add inoculum to moist seed
5. Mix thoroughly
6. Dry in **shade** for about 30 min to 1 hour
7. Plant.

Implications for extension

When it is appropriate to use inoculants and they are available they represent one of the best returns on investment for smallholders. For a very small outlay they improve the legume yield and support better biological nitrogen fixing in the soil. But the inoculant is a living organism and needs to be treated with care and attention - so good handling techniques for inoculants need to be shared and understood

4 Cropping system / agronomic approach

refers to the crops and crop sequences and the management techniques used on a particular field over a period of years so as to manage the fertility of the soil. The cultivation of the crop may sometimes be interspersed with a period of fallow. Examples of cropping systems are strip cropping, crop rotation, monocropping and intercropping.

5 Intercropping

Intercropping means growing two or more crops together on the same field. By combining crops that have different growth patterns (e.g. maize and beans), the available air, water and nutrients can be better utilized.

Intercropping can involve planting deep rooted and shallow rooted crops together.

The contribution of intercropping to soil fertility management

- With multiple crops, each with its own root pattern, water and nutrients can be absorbed from various layers and places. These resources are thus utilized more efficiently than when only one crop is grown.
- When a legume is used in intercropping, it fixes nitrogen into the soil for it to be used up by the other crop. It also suppresses weeds and thus acts as a break to pests and diseases.

It may be possible to plant two short-duration soybean intercrops with cassava before the cassava canopy closes.

4 Common examples of intercropping in Ghana include:

Maize: cowpea intercrop

Cassava: cowpea

Sorghum: groundnut (peanut)

Maize: bambara groundnut

Maize:soybean

6 Crop rotation

Crop rotation is the practice of growing a series of dissimilar types of crops in the same area in sequential seasons. Crop rotation may include 2 to 6 or more crop rotations over numerous seasons. A two crop rotation such as maize and soybeans or maize and alfalfa use legumes to fix nitrogen in the soil for utilization over a long term.

The contribution of crop rotation to soil fertility management

- It adds nutrient to the soil. Legumes of the family Fabaceae have nodules on their roots that contain nitrogen-fixing bacteria hence, alternating with cereals help fix nitrogen to the soil.
- Rotation methods that leave crop residue on top of the soil protects soil loss by erosion.

- With more soil organic matter, water infiltration and retention improves, providing increased drought resistance and decreased erosion.

Example of cereal-legume rotation is illustrated below.

Soybean to be replaced with maize next season

Maize to be replaced with soybean next season

7 Crop rotation and ISFM

Rotational benefits do not only include enhanced nitrogen supply to maize but other effects such as reduction of soil-borne diseases.

Nitorgen supply from biological nitrogen fixing can complement nitrogen fertilizer application and improve soil fertility status. Combining mineral fertilizers with organic resources may result in greater nutrient use efficiency. However, achieving this effect requires strategic management of fertilizer in terms of form, timing and placement as well as a sufficient supply of organic resources.

Site specific-adoption of ISFM principles takes into account differences in soil fertility status within fields and farms and assures more efficient use of applied mineral fertilizers and available organic resources.

The fields around the house or village are often much more fertile than the fields further away. African farmers are excellent spatial manipulators of soil fertility, creating relatively rich and fertile islands by applying both organic and mineral fertilizers to the more accessible and secure field, often at the expense of more distant fields and communal lands.

Decisions on fertilizer use and choice of cropping systems must be tailored to these differences in soil fertility and availability of organic resources.

A critical soil carbon content is required to obtain crop responses to fertilizer application. In fields with moderate carbon content, applying fertilizer accompanied by correct management strategies can considerably increase crop production. Some of the nearby fields are fertile to the extent that crops no longer respond to additional nutrients supplied through fertilizer but even in such fields, periodic application of a maintenance fertilizer dose is required to sustain yields.

Distant fields are frequently infertile, depleted and have very little soil organic matter contents. Farmers have few options – they have to rehabilitate these fields through organic resource management because crops respond poorly to chemical fertilizer application alone.

ISFM encourages smallholders to develop practical skills in the production, collection, processing and placement of organic resources. Different types of organic resources have various, and sometimes conflicting uses but practical field evaluation procedures are available to assist in their allocation.

In some cases, the activities of soil biota (plant and animal life in the soil) clearly conflict with farmers' objectives in organic resource allocation but offer other, long-term environmental benefits.

Estimating nutrient addition through the application of organic resources is more difficult than with mineral fertilizers, and their nutrient release patterns may be less predictable, but interactions between mineral and organic inputs tend to be strongly beneficial.

Smallholders must be conscious of organic resource allocation to the extent that some parts of the farm become degraded at the expense of other more accessible areas.

Finally, skilled organic resource managers should not become confused with organic farmers as sometimes occurs by development agents and donor representatives as they operate under less prescriptive management guidelines and usually include manufactured mineral fertilizers within their soil management strategies.

Chapter 5

The Concept of Integrated Soil Fertility Management (ISFM)

Summary

ISFM has been defined in a number of ways – but here an emphasis is placed on finding a way to explain ISFM that makes sense to the intended audience. When combined with general good agronomic practices the combination of improved planting material, chemical fertilizer along with the addition of organic mater, increases productivity. There are, however, constraints at the smallholder level – in terms of access to farm inputs e.g. seed, credit &labour.This section looks at short and long-term approaches to the application of basic nutrients. The issues of recycling nutrients and the impact of erosion on the nutrients in the soil. There is a review of the pros and cons of ISFM – it does increase yield (4 -6 times more production). There is a brief outline of the role of cover crops, mulching and conservation tillage as forms of agriculture that can help reduce soil erosion.The section locates the idea of ISFM along side integrated pest management and explores the impact of biological control agents in Ghana and the role of micro-organisms.

1 Definition of Integrated Soil Fertility Management (ISFM)

The Africa Soil Health Consortium defines ISFM as:

'A set of soil fertility management practices that necessarily include the use of fertilizer, organic inputs, and improved germplasm combined with the knowledge on how to adapt these practices to local conditions, aiming at optimizing agronomic use efficiency of the applied nutrients and improving crop productivity. All inputs need to be managed following sound agronomic and economic principles.'

But for farmers they define it as

Improved seeds + Fertilizer + Manure!

Good soil Good harvest Better life

Get healthier crops with ISFM

ISFM = Improved Seed + Fertilizer + Manure

Improved Seed

Use improved seed, stem cuttings, suckers and roots because they mature faster, are resistant to weed and disease attack, and are tolerant to drought. Select seeds that work well in your area and combine with fertilizer and organic matter.

Fertilizer

Applying the correct fertilizer in the right amounts, in the right way at the right time will increase the quantity and quality of your yield - especially when combined with organic matter. Using less than the recommended amounts or no fertilizer at all can result in poor yields.

Manure & other organic matter

To make fertilizer work better, add manure or other organic matter that is available in your area. (e.g. chicken dropping, cow dung, stover and compost).

For good yields combine these with good agricultural practices (correct planting time and spacing, timely weeding, not planting the same crop every season and water management)

Africa Soil Health Consortium
CABI, ICRAF Complex. P.O. Box 633-00621 Nairobi, Kenya
(t) +254-20-722 4450 (e) africa@cabi.org www.cabi.org/ashc

The goal of ISFM is to develop comprehensive solutions that are adapted for local conditions such diverse factors as weather, the presence of weeds, pests and diseases, inherent soil characteristics, history of land use and spatial differences in soil fertility.

It involves a range of soil fertility enhancing methods, such as:

- Improved crop management practices
- Integration of livestock
- Incorporation of leguminous crops
- Measures to control erosion and leaching
- Measures to improve soil organic matter maintenance.

Other terms used in literature to identify the ISFM approach include:

- Soil fertility management (SFM)
- Integrated soil nutrient management (ISNM)
- Integrated nutrient anagement (INM)
- Integrated plant nutrition systems (IPNS).

The overall objective of ISFM is to enhance the productivity and sustainability of farming systems. It provides an approach, which needs to be tailored to the characteristics of the site, and constraints faced by the farmer. This approach demands an adaptive response based on a new partnership between researchers, farmers and extension workers.

2 The context for ISFM

Over the next 40 years the population of sub-Saharan Africa is set to increase by 700 million inhabitants.

This translates into (a massive increase in the need for food, feed, fibre and fuel) in a region where many countries already import significant amounts of food.

But how can food, feed, fibre and fuel production be increased?

While it is likely that there will continue to be some further expansion in the area cultivated, there are many competing claims on land for urban development and for wilderness. Given current crop yields there is great potential to increase agricultural production through intensification of production on land already under cultivation.

Yield intensification is usually concerned with increasing the yield of crops but may also involve increasing the number of crops grown in each field each year. In addition to the sparing of land for other uses, yield intensification has benefits of increasing returns to labour (i.e. reducing the drudgery of intensive labour investment for little return), and increasing farmers' food self-sufficiency and incomes.

The bulk of sub-Saharan Africa's food requirements will continue to be produced by small-scale farmers who represent about 70% of the population in the region.

The term 'smallholder farmer' is an umbrella term that encompasses a huge diversity of types of farms within a myriad of farming systems. Distinction can be made between two types of small-scale farmers:

- Farmers engaged in the production of crop products and livestock for sale in local markets.
- Farmers engaged in agriculture either to achieve food security or as a sideline activity to supplement livelihoods based on employment or small-scale business activity.

In both farm types, improvements in soil fertility can contribute to increased yields but the appropriate approach to soil fertility improvement may be very different. For example, farmers linked into the market are usually in a stronger position to borrow money from the bank and invest in inputs (improved seed, fertilizers, agrochemicals) by comparison with farmers producing for local consumption who may not be able to borrow money to purchase inputs and are often averse to the risk of investing in agricultural inputs.

For this reason, ISFM places great emphasis on adapting proven principles of soil fertility management to the farmer's situation and goals (i.e. production for the market or for local consumption).

Improvement in agricultural productivity by small-scale farmers – the so-called 'Green Revolution' – has underpinned the economic developments that have taken place over the past 50 years in Asia. Industrial development has taken place but food security has been maintained at regional and often national levels, and small-scale farmers now benefit from expanded markets for their products in rapidly growing cities. The Green Revolution focused attention on improving productivity in lowland, and usually irrigated, rice-based systems where variability between farms is much less than the variability between farms and landscapes found in sub-Saharan Africa.

The emphasis was placed on wide-scale implementation of 'best-bet' technologies that could be implemented effectively across large areas. As we shall see, farming system development in sub-Saharan Africa requires very different technologies and approaches to productivity improvement to those used successfully in the predominantly irrigated farming systems in Asia. Nevertheless, some features are common to both regions, particularly with regard to the role of the state as:

- A primary driver of agricultural productivity improvement in small-scale farms;
- A source of finance for infrastructure and institutions required to better integrate farmers into markets for inputs (i.e. fertilizers, seeds, agrochemicals and credit) and outputs; and
- A source of research and extension leading to the dissemination of information on appropriate technologies for soil fertility management to a diverse range of farmers.

ISFM has the greatest potential for impact in sub-Saharan Africa in areas where:

- There is a need for crop intensification due to high and increasing population.
- Farmers have access to markets for their products.

3 The merits of ISFM

Advantages of integrated soil fertility management	Disadvantages of soil fertility management
Increased yields (2.33 – 3.92 tons per hectare, UDS/AGRA soil health extension project report; Aug 2012-Jan 2013)	The low nutrient concentration of many organic fertilizer means that a very large amount of material must be transported and applied to attain any reasonable yield.
Waste lands are converted into productive land.	Applying inorganic ferilizers can expensive which may require credit and increased financial risk for farmers.
Integration of cereals and legumes allows farmers to continuously achieve high yields on the same land for many years, eliminating the need for clearing new lands, which is rarely available.	High levels of labour invested in transport and application or organic and inorganic inputs.
Makes better use of organic materials available on-farm to build up soil organic matter.	Only really feasible where the crops grown can fetch a good price, such as for vegetables around a major town or where crops can be stored to take advantage of post harvest price increases.

Incorporates the use of legumes and inoculants to increase nitrogen levels - which is green and low in cost.	Programmes which rapidly increase productivity in an area without concomitant market development - can see price crash and harvests become worthless.
Farmers share knowledge by working in groups.	
Improved soil moisture and nutrient retention.	
Ghana has access to ready markets at home and in neighbouring countries.	

Implications for extension

When trying to encourage the uptake of ISFM practices it is important to identify suitable entry points for smallholders. These entry points should produce a large return for farmers relative to the inputs or changes in production practices. This audit process is sometimes called farming systems analysis (FSA). This will determine:

- The point of the farming system which is the priority for improvement.
- The impact that this should have on the farming family.

Change guru John Kotter has good advice on how to make change stick. This includes having a clear vision of the change you are trying to make and having incremental progress which is celebrated as part of the push towards larger change in working practices and embedding the change.

Remember ISFM will increase yields - so it is essential that adequate thought is given to the markets that will be accessed. Local markets can be overwhelmed by localized productivity increases and market prices are driven by demand and supply. So ISFM has to increase livelihoods as wells as productivity.

4 Replenishing the pool of soil nutrients

Traditionally fertility regeneration/replenishment was achieved through long fallow periods. Areas were rested or left to fallow for periods up to 10 years - then often slashed, burned and tilled to make them ready for cultivation.

Increased population has either wiped away or very much reduced these fallow periods. And increased urbanization has created a pressure for more settled agriculture that can be permanently productive.

Replenishing the pool of soil nutrient is mainly achieved by addition of external inputs. Adding nutrients is done to achieve two main purposes, namely, short term and long term purposes.:

1. **Short-term purpose** in soil fertility management, farmers may add mineral fertilizers, apply farmyard manure or grow a green manure crop.
2. **Long-term purpose** include fallowing, application of organic matter with a high carbon/nitrogen content, application of one-time high doses of inorganic phosphorus or phosphate rock.

Time of application

Pre-planting	Planting	Standing crops
Organic matter, like farmyard manure, compost and leaf mould are incorporated into the soil well in advance before sowing/planting of vegetable crops. When this is done these manures get mixed properly in the soil and start rotting to release nutrients when crop plants are in need of them.	**Chemical fertilizers** are applied as a basal dose and also in the form of top dressing. The basal is applied just one day before sowing or planting and mixed or drilled into the soil. Care is taken for the presence of sufficient soil moisture.	**Chemical fertilizers** are used for top dressing of fertilizer, particularly nitrogenous fertilizers is done 15 -21 days after sowing/planting. This is the time when plants get established. **Foliar feeding** of nitrogen and micronutrients is when plants start showing deficiency symptoms.

5 Adding nitrogen (N)

Although one of the most abundant elements on earth, nitrogen (N) deficiency is probably the most common nutritional problem affecting plants worldwide.

Most plants take nitrogen (N) from the soil continuously throughout their lives, and demand usually increases as plant size increases.

Building soil nitrogen is made difficult because of the dynamic nature of nitrogen cycling in the soil.

Nitrogen is stored in three main forms

- Mineral nitrogen (ammonium NH_4 and nitrate NO_3)
- Nitrogen in relatively labile soil organic matter (ie. Nitrogen stored in OM that is still undergoing change/decomposition)
- Nitrogen in a more stable form of soil organic matter (Nitrogen stored in humus).

Nitrogen is sometimes referred to as the "nutrient that doesn't want to stay put". Nitrate is highly mobile and easily lost by leaching or by denitrification (lost as gases NO, N_2O and N_2). Substantial losses of NH_4-N can also occur through volatilization (gaseous losses as NH3).

Labile - easily altered

Legumes may be used to build soil nitrogen. However, in many legumes the nitrogen is fixed in the grain. The growing of soybeans, for instance, may even result in a net removal of nitrogen from the field.

When soils are cultivated continuously it may be impossible to build up soil nitrogen. Even when the store of nitrogen is replenished, continued use of crop sequences and intercrops with grain legumes and green manures, better integration of crops and livestock and optimal use of mineral fertilizers are essential to ensure improved yields are maintained.

Opportunities for the build-up of nitrogen capital associated with organic matter may be limited in sandy soils with weak microbial activity and poor structural protective capacity of the soil, This is especially the case in lowland areas where temperatures are high.

6 Adding phosphorus (P)

Phosphorus is lost though crop-harvest removals and soil erosion.

The phosphorus content of plant residues and manure is not usually sufficient to meet crop requirements. Adding phosphorus fertilizer is usually necessary to overcome phosphorus depletion.

Soil phosphorus can be replenished with soluble phosphorus fertilizers, direct application of sufficiently reactive phosphate rock, or the combination of soluble phosphorus fertilizer and phosphate rock.

Highly phosphorus-absorbing soils may require high levels of additional phosphorus before a response is achieved, although the residual benefits of such applications are potentially high. That means that the phosphorous will be available to the plants in the future.

By contrast, in sandier soils, lower applications may be given. However, the residual benefits will probably be limited, due to the combined effects of leaching of inorganic phosphorus and the limited existing fraction of organic phosphorus.

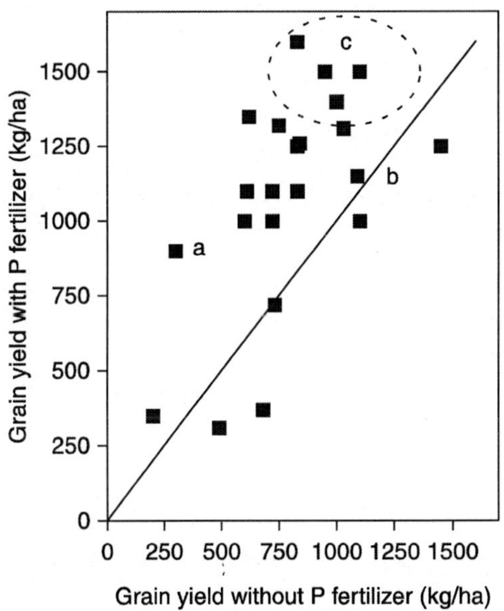

The response rate of different soyabean varieties to phosphorous fertilizer

The suitability of phosphate rock for direct application to soil depends upon its mineralogy and reactivity, soil properties and climate, crop. The economics of the use associated with the phosphate rock must also be considered.

Dissolution of phosphate rock requires low pH, low soil exchangeable calcium (Ca) and low soil solution phosphorus concentration. So a more acidic rhizosphere will enhance the plants ability to access the dissolution of phosphate rock. A low soil solution phosphorus concentration resulting from high phosphorus sorption may limit plant growth. Combined use of phosphorus fertilizers and legumes can result in synergism where nitrogen fixation can be enhanced because of reduction of phosphorous deficiency.

Sorption - the process in which one substance takes up or holds another; adsorption or absorption.

The **rhizosphere** is the narrow region of soil that is directly influenced by root secretions and associated soil microorganisms.[2] Soil which is not part of the rhizosphere is known as bulk soil. The rhizosphere contains many bacteria that feed on sloughed-off plant cells, termed *rhizodeposition*, and the proteins and sugars released by roots. Protozoa and nematodes that graze on bacteria are also more abundant in the rhizosphere. Thus, much of the nutrient cycling and disease suppression needed by plants occurs immediately adjacent to roots.

7 Maximizing on-farm recycling of nutrients

Maximizing on-farm recycling of nutrients requires effective crop residue management. The key is to ensure that adequate residue exist throughout the cropping season.

Crop residue could be used as long term soil cover (mulch see Chapter 3 section 1) to protect the surface of the soil. Residue could also be allowed to decompose to release nutrients into the soil.

Alternatively crop residues can be used for **compost**preparation which can be incorporated into the soil to enhance fertility.

8 How soil nutrients are lost to the environment

Summary of terms

Leaching

Refers to the loss of water-soluble plant nutrients from the soil, due to rain and irrigation. Adjustments to soil structure, crop planting, type and application rates of fertilizers, and other factors can be employed avoid excessive nutrient loss.

Volatilization

Volatilization is the process whereby a dissolved sample is vaporised (turns into vapour).In the case of urea, hydrolysis converts the urea to ammonia, and if the urea is not incorporated, the ammonia is lost to the air.

Conditions favouring high volatilization potential are:

- High soil temperatures
- Moist conditions, followed by rapid drying
- Windy conditions

- High soil pH (>pH 7.5)
- High lime content in surface soil
- Coarse soil texture (sandy)
- Low organic matter content
- High amounts of surface residue preventing the incorporation of the input into the soil (e.g. zero tillage)
- As a rule, nitrogen fertilizers especially urea should be incorporated into the soil after application (not surface applied) to prevent its loss to the atmosphere.

Fixation

Nutrients such as phosphorus are easily fixed, making them insoluble.

Although the nutrients are not lost from the soil, fixed nutrients can only become useful to crops when they are able to dissolve in soil moisture and this may take a long time.

Placement is critical to *prevent* phosphorus *fixation* and nitrogen losses to immobilization

The loss of nutrients from the soil can be reduced using erosion control methods below.

Implications for extension

There isa range of opportunities for introducing nutrients to the soil - many of them are constrained for smallholders - so it is essential that extension officers understand and can explain the options in ways smallholders can understand.

It is important to understand the level of impact that legumes can have the soil nitrogen content. The use of inoculants is a low-cost way to boost nodulation and nitrogen fixing. But it is also important to leave - or return as much of the legume plant matter to the fields.

Legumes can be harvested by cutting them not pulling them - minimizing the soil disturbance and retaining some of the nitrogen. All other plant matter - excluding the grain crop should be returned to the soil - even if it has passed through the livestock to get there.

Being clear about the soil type is important too - this will impact considerably on the way that nitrogen behaves.

9 The impact of soil erosion on nutrients

Soil erosion is the physical removal of nutrient rich, high organic matter top soil by water (runoff) or wind leading to subsoil exposure.

It reduces the productive capacity of the soil. Nutrients including phosphorus (P) are carried away in eroded sediments.

Recent studies indicate that annual erosion losses in low-input production systems in sub-Saharan Africa are about:

- 10kg nitrogen /hectare
- 2kg phosphorus/hectare
- 6 kg potassium /hectare.

The photograph below shows severe erosion on land ploughed with tractor.

Severe erosion on a tractor ploughed field. The gullies have developed as a result of erosion.

10 Methods of controlling soil nutrient loss and erosion

Development control strategies must consider the mechanism of soil erosion including slope and slope length to reduce overland flow velocity.

11 Agronomic/ biological erosion control approach

- **Cover cropping** (see chapter 4 section 11) is very important in reducing erosion. The major role of vegetation is to intercept rainfall and reduce the velocity and kinetic energy with which they impact the soil. Vegetation also increases surface roughness of the soil and slows down flow velocity, and hence reduces further detachment and transport of eroded material. It must be noted that vegetation that spreads is more effective in reducing erosion than erect type and the effect of plant cover changes with the stage of the crop.

- **Mulching** (chapter 4 section 1) is the covering of soil with crop residue such as straw, maize stalks, stubble etc. This cover protects the soil from rain splash and reduces the velocity of overland flow.

Nutrient losses may be greater in high-input systems, or where rainfall is very high.

Erosion and runoff can also be reduced by covering the soil with a mulch layer of living or dead biomass. Soil mulch reduces water speed, avoids crust formation and improves soil porosity and infiltration rates. Even a relatively thin layer of mulch provides a significant increase in water infiltration.

Studies have shown that the application of 2 tonnes of straw/ hectare led to a 60% reduction in runoff and a 90% reduction in erosion.

With 6 tonnes of straw/ hectare, runoff was reduced by 90% and erosion levels were reduced to zero. Leaving straw in the field leads also to significant reduction in soil losses due to wind erosion.

Mulching

Use of Herbicide

- **Conservation tillage** (see Chapter 4 - 1b) refers to any tillage that leaves more than 40% residue on the surface of the soil. Some types of conservation tillage include minimum tillage, zero tillage etc. The main issue in conservation tillage is that seed is drilled directly through the stubble and weeds are controlled by use of herbicide. Other agronomic approaches include crop rotation shifting cultivation etc.

12 Engineering approach

The second approach to erosion control is the engineering approach. These include practices such as contour ploughing terracing and bunding.

Water barriers, such as grass strips, and stone rows are effective options to reduce erosion and to keep applied fertilizer and manure in place.

- Zai technique

Soil preparation methods may also be efficient in increasing infiltration and reducing runoff. The so-called 'Zaï' technique is an effective technique to deal with surface crusting: small pits are dug in the soil and small amounts of mineral and/or organic fertilizers are added. Improving the soil organic matter content will generally reduce the susceptibility of the soil to form surface crusts and improves soil structure and water holding capacity.

13 Improving the efficiency of external inputs

Application of organic resources of animal or plant origin in combination with mineral inputs can maximize input use efficiency. While important, adding nutrients should not be considered only in terms of just single elements.

Soil fertility is the ability of soil to supply plant nutrients in adequate and balanced amounts. Thus once a soil can supply nutrients in the right proportions, then the soil is said to be fertile with respect to the crop that can absorb those nutrients. Sometimes there is confusion between soil fertility and productivity. Soil fertility deals with only the chemical properties of the soil. It deals with the pH, CEC, AEC, the cations and anions, which are absorbed by plants, fertilizers etc.

Soil productivity on the other hand is the ability of a soil to yield crops. Soil productivity involves the physical, chemical, microbiological, biochemical and all other properties of a soil. Soil fertility is thus a component of soil productivity.

A soil may therefore be fertile but not productive. For example the Akuse Series, found in and around Akuse and Kpong in Ghana and the Siarre series found in lowland of the Upper West and East regions of Ghana contain heavy clays, and are very fertile but not productive. These soils, which contain expanding 2:1 clays (smectite eg. montmorillonite), have very high CEC and are rich in basic cations. However, because they are sticky and expand when wet and very hard and crack when dry, they are very difficult to manipulate. Productivity of such soils though very fertile is low.

14 Protection of beneficial soil micro-organisms

Beneficial soil-borne microorganisms, such as plant growth promoting rhizobacteria and mycorrhizal fungi, can improve plant performance by inducing systemic defense responses that confer broad-spectrum resistance to plant pathogens and even insect herbivores.

Soil microbes act as a type of probiotic, or immune system, for soil, effectively preventing plant diseases from developing. There is proof that synthetic pesticides and herbicides are completely unnecessary when the right balance of natural microbes are present and flourishing in soil.

In 2012 The Environment Protection Agency in Ghana approved the first biological control agents.

15 Integrated pest management

In 2011 four biological control agents were now registered and approved for use by the Environmental Protection Agency of Ghana.

The first four biological control agents are: **Campaign** (trade name) is a fungus (*Metarhizium anisopliae*) for the control of mealy bugs in papaya. This product has been commercially launched and has already had its first sales. Campaign is likely to also be effective against **thrips, white fly and caterpillars.**

Sustain (trade name) is a fungus (*Trichoderma asperellum*) which has been registered for the control of *Phytophthora* in pineapples, but which will also be registered for the control of **black pod disease in cocoa**. This product has been commercially launched in June 2011.

Heltec and **Plutec** (trade names) are two baculoviruses (HaSNPV and PlxyGV) that are undergoing efficacy trials for the control of bollworm in tomatoes and against diamondback moth which is a major pest of **cabbages and other *Brassica*.** Encouragingly, toxicological trials of the baculoviruses being carried out in India have demonstrated no toxic effects. These products are awaiting their commercial launch.

Background Ghana exports 90,000 tonnes of fresh produce per year to the European Union (EU). The cocoa sector in Ghana, which has 800,000 small-scale growers and produces 379,000 tonnes per year, is also under pressure to adopt integrated pest management (IPM). To maintain or grow its EU market share, Ghana needs to reduce its pesticide use.

Chapter 6

Organic Resource Management within ISFM

Summary

The use of organic resources in modern agriculture implies the production of crops through organic farming without doing any damage to the environment. Synthetic chemicals are never used in organic farming. This avoids having much pressure on nature's non-renewable resources used in farming activities. In this unit, we shall discuss many organic resources, their quality and management within ISFM.

Sources of plant nutrients include residual soil nutrients, crop residues, green manures, animal manures and biosolids, biologically-fixed nitrogen and manufactured fertilizers. Crops use plant nutrient from all sources but the nutrient elements must be transformed into the ionic forms before being taken by plants. The amount of nutrients provided by different sources varies between and within agro-ecosystems. Integrated plant nutrient management identifies and uses all available sources of plant nutrients

1 Organic resource management

There is abundance of organic resources derived from both cultivated and natural lands.These resources are, however, under-utilized by smallholders within the context of ISFM.

The uptake of organic resources as soil nutrients sources is limited by

- Their alternative uses as fuel
- Their use as feed and fodder
- Fibre
- The labour required to collect and process these materials.

Plant residues and livestock manures decompose rapidly in moist and warm climates, causing nutrient release to be poorly timed with crop demand. This suggesting that the timing and placement of organic resources must be carefully considered.

In many cases, organic resources most available to farmers have low nutrient concentrations with limited potential to improve crop yields when applied as the sole source of nutrients.

In contrast, has been widely tested in the tropics for its potential to sustain adequate food production under low external inputs. Within smallholder communities, the demand for animal manure is usually greater than its limited supply and in pastoral areas with substantial livestock, free grazing poses difficulties in collecting and transporting this important organic resource.

These difficulties must not preclude the use of organic materials as inputs to soil but rather require that they be utilized in more labour efficient and cost effective ways.

Meeting even modest food production and rural development objectives in sub-Saharan African, demands strategic use of limited available resources.

ISFM interventions therefore aim to increase crop production through improving the agronomic efficiency of applied nutrient inputs. This approach necessarily involves the use of farmer-available organic resources and appropriate agronomic practices. These must be adjusted to local conditions as a means of both delivering nutrients and improving the efficiency of applied mineral fertilizers.

These practices require informed actions by land managers because they must be adjusted to site-specific conditions. For this reason, ISFM is best achieved through the application of flexible principles that increase the availability of organic resource to farmers and makes best use of items.

2 Organic resource quality

The mineralization (release of nutrients), during decomposition and transformation of organic materials into soil organic matter, is regulated by the interaction between the decomposition of various organic materials. This involves soils, and surface organisms, the physical environment and the chemical characteristics of a given substrate.

3 Organic resources available to farmers

The success of agriculture, especially organic agriculture, depends upon the development and integration of various farm activities in such a way that availability of the organic resources does not become a constraint.

Suggested exercise define the difference between organic resources and inorganic resources define the difference between organic resources and organic farming

4 Organic sources of plant nutrient value available to farmers

This section discusses issues on availability of organic resources to farmers.

Inoculants (see chapter 4)

Green manure (see chapter 4)

Crop residues (see chapter 4)

Cover crops (see chapter 4)

Manure (see chapter 4)

5 Agro-industrial by-products

Agro-industrial wastes result from the first-step processing of agricultural commodities. These by-products are potentially important sources of organic materials but pose difficulties for more distant smallholder famers to access. In many cases, the agricultural raw materials are produced by small-scale out-grower farmers, transported to processing plants and then used by the central processing facility or nuclear plant and not returned to the fields and farmers of origin. Examples of these products include sugarcane bagasse, coffee husks, tea powder, rice husks and coconut husks.

Locally available agro-industry by products

Agro-industry by-products that are likely to be available more locally include:

- **Bran** (produced by millers that grind cereals into flour). This nutritious material is generally fed to livestock.

Other organic sources of plant nutrient include aquatic weeds. These are organic material that are available to farmers near water bodies.

Water hyacinth is an aggressive aquatic weed that has invaded many waterways of sub-Saharan Africa and must be periodically cleaned from harbours, dams and canals. Pit composting of water hyacinth reduces moisture content and increases its nitrogen concentration on a dry weight basis. This transformation greatly improves the economics of transporting these materials to farmers' fields.

6 Biosolid

Biosolids are the residual solids from treatment of municipal wastewater. They can be recycled and provide significant quantities of plant nutrients as well as organic matter.

Nutrients in biosolids vary in forms, depending on the source, treatment, storage and handling processes. Nutrient availability depends on the particular biosoild and the local environment, just as with manures.

Generally, biosolids are analyzed for their content of organic and inorganic nitrgen, total phosphorous, trace elements and toxic metals (such as mercury and arsenic) prior to land application.

Plant-available nitrogen values for biosolids and are derived from the nitrogen analysis, method of application, time of application and local climatic situation (which influences mineralization).

Phosphorus release from organic P forms is difficult to predict, but application rates to supply adequate N for crop growth generally supply more P than most crops need and, biosolids at rates to supply all crop N needs can increase soil P concentrates to levels that may produce environmental concerns for P runoff to surface waters.

The agricultural use of biosolids is supported by the United Nations Environmental Programme (UNEP) - they identify the following steps for proper use of this resource:

- The composition of biosolids must be carefully evaluated and closely monitored in order to assure accepted quality and to calculate proper application rates;
- Chemical and biological properties must be monitored regularly, including trace elements, pathogens and odours to ensure protection of crops, land and the individuals producing and utilizing the crops;
- Knowledge and understanding of physical, chemical and biological conditions of soils and crops are needed to determine appropriate rates, methods and timing of applications;

- Laboratories and research programmes supporting the utilization and monitoring must consistently do high-quality work in order to ensure the integrity of the biosolids utilization plan. Specifically, laboratories should have well established quality assurance procedures and meet ISO standards

7 Processing and application of organic resources

There are a number of processes involved in transforming some of the organic resources and turning their plant nutrients into available forms for plant assimilation.

Collection and storage

Organic resources may be either gathered and deployed, or collected and stored for use in the growing seasons when crop nutrient is in demand.

Direct deployment includes the establishment of trash lines and mulches from crop residues and chopping and incorporation of green manures.

Collected and stored organic matter can be bulked for latter us. Practices that are particularly well suited to crop residues and animal manures.

Stored organic resource include piling crop residues as livestock feed (FYM) during the dry season, heaping manures and the production of compost. It is important to protect stored organic materials from excess rainfall, runoff and leaching. This goal may be achieved by covering organic heaps with tarpaulins or placing them in sheds. In many cases, organic materials must be well dried in the field and well aerated during storage to prevent further decomposition.

Pre-plant incorporation

One of the most practicable uses of organic matter is to apply them during land preparation. This strategy combines organic input management with field operations such as tillage and fertilizer application.

Green manure should be applied several weeks before tillage by several because vegetative cover must be chopped or grazed in order to reduce its bulk. This particularly true when hand hoeing or draft animals are to be used for ploughing.

Nutrient concentration for organic material

One important field operation is the spreading of stored organic materials such as animal manures and composts. One approach to allocating these materials within the field is to calculate the distance necessary between piles of known nutrient concentration that are required to obtain a targeted amount of nutrients.

These piles are often distributed by wheelbarrow or in bags containing approximately 25kg of organic inputs. Depending on the nutrient concentration and the targeted nutrient addition, these piles are spaced between 4m and 12m apart.

Smallholders must learn to calibrate the placement of organic piles in the field, and their subsequent spreading to targeted rates of nutrients application. In addition, spreading and incorporation may be combined with pre-plant application of mineral fertilizers to simplify field operations.

Caution must be exercised in applying low quality materials, even in conjunction with mineral fertilizers. Organic inputs extremely low in nutrients and high in lignin and polyphenols must not be incorporated into the soil as these inputs will likely result in immobilization of soil nutrients and applied fertilizers. That is they will harm the fertility of the soil so these materials are best applied as surface mulches.

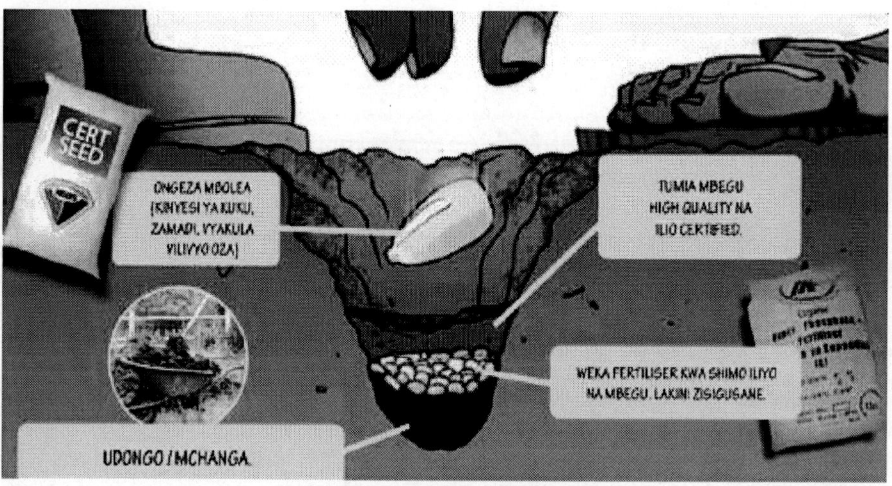

Manure can be microdosed

Microdosing: The application of small quantities of fertilizer (about 6 grams, a full bottle cap or a three-finger pinch) or manure in the hole where the seed is planted (at the time of planting). That translates to about 67 pounds of fertilizer for every 2.5 acres. This technique uses only about one-tenth of the amount typically used on wheat, and one-twentieth of the amount used on corn in the USA.

Fertilizer Microdosing Boosting Production in Unproductive Lands

Composting

Composting is a practical means of bulking organic resources and concentrating their nutrients.

The process of decomposition or composting must be controlled, particularly through the choices of substrate, moisture content and aeration.

Composting is characterized by a period of rapid decomposition and temperature accumulation followed by cooler, slower decay of the remaining organic substrate.

The rate of decomposition can be increased by stacking the materials in a pile to a height of 1 -1.5 m. However, taller stacks must be more regularly turned to facilitate rapid decomposition and prevent the formation of unwanted anaerobic by-products.

The most important physical properties to composting are particle size and moisture content.

Particle size affects oxygen movement into and within the pile, as well as microbial and enzyme access to the substrate. A proper balance in the particle size should be maintained. If too large, the organic materials should be mixed with a bulking agent such as wood chips or bagasse (the fiber left over after the juice has been squeezed out of sugarcane stalks).

The optimum moisture content for composting is 40-60 % as excess water interferes with oxygen. Too little water hinders diffusion of soluble molecules and microbial activity.

The relative quality and quantity of the organic residues affects the rate of compositing and the characteristics of the finished products. For example, when the carbon to nitrogen ratio (C/N) of the organic matter is about 25%, transformation of the organic material proceeds rapidly with a high degree of efficiency of nitrogen assimilation into the microbial biomass.

A narrower C/N ratio (>40%) may promote immobilization of available nitrogen in the compost, slowing the rate of decomposition. Therefore, addition of mineral nitrogen (and phosphorous) can enhance more rapid decomposition and enrichment of the low quality residues.

9 Fortified compost

Low quality organic materials such as maize stover or wheat straw with a C/N ratio of 80:1 are suitable for preparing fortified compost.

Instructions on how to make fortified compost

Chop crop residues to 30-45cm length (they should not be longer than your two feet in front of each other)this will increase the increase their surface area.

The chopped material is placed in layers of 30cm (the size of your foot) high by 2m (about 3 paces) wide by 25m (about 30 paces) long = 500kg in each layer - the aim is to build up 5 layers.

At every 30cm layer, evenly broadcast 4kg DAP (or any other nitrogen-bearing fertilizer) for fortification lowering the C/N ratio from about 80 to12.

Spread 1kg of farmyard manure or sugar mill filter mud, to serve as a microbial activator.

Apply 20 litres of water to enhance dissolution of fertilizers and to moisten the stover for microbial activity.

This process continues stepwise until the stack reaches 1.5 m (5 layers)

10 Worms and vermicompost

An alternative approach to compositing involves epigeic earthworms, ones that live within and consume plant debries. These worms are domesticated and, when fed a variety of organic materials, they produce vermicompost. These composts are rich in plant nutrients and have excellent physical properties.

Useful vermicomposting species include:

- **African night crawler (*Eudriluseugeniae*)**
- **The tiger worm (*Eisenia foetide*)**is the most commonly utilized species in commercial vermiculture and waste reduction. The species colonizes many organic wastes and is active in a wide temperature and moisture rang. The worms are tough ,readily handled, and survive in mixed species cultures.
- ***Perionyx excavates*** is another species well adapted to vermicomposting in the tropics that is profile and easy to handle but it cannot tolerate temperature below 5°C.

Instructions on how to make vermicompost

Vermicompost is typically produced in raised beds that are well aerated, moist and covered.

Prepare a bed with floor and walls 20-30 cm in height and line it with chicken wire for better handling and aeration Fill the bed with a 10 to 15 cm layer of coarse organic materials.

Place another 5 to 10 cm layer of animal or green manure on top of the coarse material. The material must not contain poultry manure as uric acid is harmful to worms. Mix some of the finer material into the coarse layer. Moisten the organic materials prior to the introduction of the worms. Fresh materials need little watering while dried materials may require as much as 30 litres m^2.

Release the earthworms into the moist bed. Apply 200 to 300 adult worms per one square meter of compost bed. Avoid handling individual worms, rather place small handfuls of material rich in earthworms (clusters) into holes spaced about 50cms (an arms length) apart.

Cover the bed with large leaves such as banana dark polythene plastic. Frequently inspect the bed during vermicomposting for moisture and the presence of predators. Ants will usually leave the bed if the underlying chicken wire is violently and repeatedly shaken.

Organic materials may be applied to the bed regularly as additional layers or in discrete locations. A common practice is to provide organic wastes frequently by burying the organic material in a different location within the bed on which the crops have been planted.

Vermicompost is ready after 2-4 months. Additional feeding prolongs the vermicomposting process but yields larger amounts of vermicompost. Withhold feed about 3 weeks before the vermicompost is collected to obtain 'cleaner' finished compost.

When the vermicompost is ready, worms are harvested and compost processed. Place a fine feed material on the bed prior to vermicompost harvesting to facilitate the collection of worms. Wheat bran, brewers' waste or fresh cattle manure are particularly good feeds that lure earthworms, collected worms may also be fed to fish and poultry.

Spread vermicompost in the sun to collect other clusters of worms by hand as the vermicompost dries. Once worms are collected, the vermicomposting cycle may be repeated. The finished vermicompost is uniform, dark and fine textured.

Vermicompost is best used as the main ingredient in a seedling or potting medium after passing it through a 5-10mm mesh. A typical nutrient content from a manure-based vermicompost using E. foetida is 1.93 %N, 0.26%P and 2.64%K.

11 Combined application of mineral fertilizer and organic resources

The combined application of organic resources and mineral fertilizers is increasingly gainingrecognition as a viable approach to address soil fertility decline in sub-Saharan Africa.

It isproperly called the integrated nutrient management system and is characterized by reduced inputofchemical fertilizers and combined use of chemical fertilizers with organic materials such asanimal manures, crop residues, green manure and composts.

Within the context of ISFM it is important to combine organic and mineral sources of nutrients to obtain the full advantages of both sources.

Combining mineral fertilizer with organic inputs can substantially improve the agronomic efficiency of the nutrient use compared to the same amount of nutrients applied through either mineral or organic source alone.

Combined application results in improved agronomic efficiency for a number of reasons.

Common mineral fertilizers lack the minor nutrients essential for crop growth. Organic resources contain these, but to meet the crop's major nutrient requirement (N, P and K) with organic matter alone would require application rates of around 10 tons of dry matter per hectare. However, use efficiency of nutrients applied through organic materials alone is often low. Combining both sources ensures the supply of all nutrients in suitable quantities and proportions.

Second, a combination of inorganic and organic nutrient sources results in a general improvement in soil fertility status. An increased soil organic matter content enables improved nutrient retention, turnover and availability. Particularly P availability is enhanced by organic residue application.

Organic amendments also counteract soil acidity and alluminium toxicity. The physical soil structure is improved, leading to reduced erosion, enhanced water infiltration and storage and improved root development.

Practices to enable efficient fertilizer use do not necessary require that the organic resources be applied at the same time. In addition, soil organic matter can be increased within the system through mineral fertilizer alone when their use results in much greater root biomass and return of crop residues. An example is the inclusion of a promiscuous, high-biomass yielding soybean into maize-based systems. But in ISFM terms the soyabean is seen as an organic input.

On P - fixing soils, P addition is essential to stimulate nitrogen fixation by the legumes and enable sufficient biomass production.

12 Crop rotation

Rotational benefits do not only include enhanced nitrogen supply to maize but other effects such as reduction of soil-borne diseases. Nitorgen supply from biological nitrogen fixing can complement nitrogen fertilizer application and improve soil fertility status.

Combining mineral fertilizers with organic resources may result in greater nutrient use efficiency. However, achieving this effect requires strategic management of fertilizer in terms of form, timing and placement as well as a sufficient supply of organic resources.

Site specific-adoption of ISFM principles takes into account differences in soil fertility status within fields and farms and assures more efficient use of applied mineral fertilizers and available organic resources.

The fields around the house or village are often much more fertile than the fields further away. African farmers are excellent spatial manipulators of soil fertility, creating relatively rich and fertile islands by applying both organic and mineral fertilizers to the more accessible and secure field, often at the expense of more distant fields and communal lands.

Decisions on fertilizer use and choice of cropping systems must be tailored to these differences in soil fertility and availability of organic resources.

A critical soil C content is required to obtain crop responses to fertilizer application. In fields with moderate C content, applying fertilizer accompanied by correct management strategies can considerably increase crop production. Some of the nearby fields are fertile to the extent that crops no longer respond to additional nutrients supplied through fertilizer but even in such fields, periodic application of a maintenance fertilizer dose is required to sustain yields.

Distant fields are frequently infertile, depleted and have severely depressed soil organic matter contents. Farmers have have few options - they have to rehabilitate these fields through organic resource management because crops respond poorly to fertilizer application alone.

ISFM encourages smallholders to develop practical skills in the production, collection, processing and placement of organic resources.

Different types of organic resources have various, and sometimes conflicting uses but practical field evaluation procedures are available to assist in their allocation.

In some cases, the activities of soil biota clearly conflict with farmers' objectives in organic resource allocation but offer other, long-term environmental benefits. Soil biota refers to all the organisms that spend a significant portion of their life cycle within a soil profile, or at the soil-litter interface. These organisms include earthworms, nematodes, protozoa, fungi and bacteria.

Estimating nutrient addition through the application of organic resources is more difficult than with mineral fertilizers, and their nutrient release patterns may be less predictable, but interactions between mineral and organic inputs tend to be strongly beneficial.

Smallholders must be conscious of organic resource allocation to the extent that some parts of the farm become degraded at the expense of other more accessible areas.

Finally, killed organic resource managers should not become confused with organic farmers as sometimes occurs by development agents and donor representatives as they operate under less prescriptive management guidelines and usually include manufactured mineral fertilizers within their soil management strategies.

Chapter 7
Mineral Fertilizer Management within ISFM

Summary

Fertilizers are materials when applied replenish lost soil nutrients on farmer's field. Without the use of fertilizers farmers would find it extremely difficult to increase their production to meet the ever-growing mouths. The purpose of this session is to discuss what a fertilizer is, and the different terminologies used in fertilizer technology.

1 What is a fertilizer?

A fertilizer is any material, organic or inorganic, material or synthetic, that supplies plants with the necessary nutrients for plant growth and optimum yield. Fertilizer allows farmers to supplement the nutrients which are already present in the soil and are able to match the supply of nutrients with the needs of the crops. Fertilizer usage by farmers is the only means by which farmers can meet their crop's exact requirements in-order to ensure sustainable productivity.

2 Fertilizer terminology

Terminologies used in fertilizer technology include the following:

A ***fertilizer*** is any substance that is added to the soil to supply those elements required in the nutrition of plants.

A ***fertilizer material*** **or** ***carrier*** is any substance that contains one or more of the essential elements.

A ***mixed fertilizer*** is a mechanical or chemical combination of two or more fertilizer materials and which contains two or more essential elements.

A ***complete fertilizer*** contains the three major plant-nutrient elements-nitrogen, phosphorus, and potassium.

Fertilizer grade refers to the minimum guarantee of the plant-nutrient content in terms of total nitrogen, available phosphorus pentoxide, and soluble potassium oxide (6-24-24 for example).

Fertilizer ratio refers to the relative percentages of nitrogen, phosphorus pentoxide, and potassium oxide (a 6-24-24 grade has a 1-4-4 ratio).

An ***acid-forming fertilizer*** is one capable of increasing the acidity of the soil, which is derived principally from the nitrification of ammonium salts by soil bacteria.

A ***basic fertilizer*** is capable of decreasing the acidity of the soil.

A ***nonacid-forming or neutral fertilizer*** is one that is guaranteed to leave neither an acidic nor a basic residue in the soil.

A ***filler*** is make-weight material added to a mixed fertilizer of fertilizer material to make up the difference between the weight of the added ingredients required to supply the plant nutrients in a ton of a given analysis

Dry bulk blending is the process of mechanically mixing solid fertilizermaterial.

Clear liquid fertilizer is one in which the NPK and other materials are completely dissolved.

3 Types of fertilizers

Soil amendments are made by adding fertilizers to the soil but there are different types of Fertilizers, namely, organic and inorganic or chemical fertilizers. However, an organic fertilizer is not often used; rather the term organic manure is preferable.

1. Organic manure

Organic matter may be defined as materials which are organic in nature, bulky and originally concentrated and capable of supplying plant nutrients and improving soil physical environment, having no definite chemical composition, with low analytical value, produced from animal, plant and other organic wastes and by products.

Organic matter may include materials such as cow manure, bone meal, organic compost, well rotten farmyard manure, green manure crops, etc. Organic matter are often the most convenient forms of fertilizers; they are safe and easily available.

Vegetation material called mulch, such as hay, peat moss, leaves, grass, bark, wood chips, seed hulls, and corn husks all help to aerate the soil, insulate the ground against temperature change and add needed nutrients.

2. Inorganic Fertilizers

An inorganic fertilizer, which is also referred to as chemical fertilizer is primarily derived from chemical compounds such as ammonium nitrate, ammonium phosphates and potassium chloride and is made-up of different formulations to suit a variety of specified uses. Chemical fertilizers usually come in either granular or powder form in bags and boxes. They may also be in liquid formulations in bottles, gallons and other containers.

Advantages of inorganic fertilizers

The advantages of inorganic or chemical fertilizers include;

- They contain higher and accurate amount of nitrogen which promotes protein and chlorophyll formation and encourages growth of stems and leaves.
- They supply higher amount of phosphorus resulting in production of more flowers, larger fruits and healthier roots and tubers.
- They also contain potassium from potash which fosters protein development and thickens stems and leaves.
- They release nitrogen rapidly.
- They provide accurate sources of plant nutrients.

Disadvantages of inorganic fertilizers

- Inorganic fertilizers can burn your plants and distort the quality of your soil leading to cadmium poisoning if used carelessly.
- Strict watering schedules have to be adopted when using inorganic fertilizers in order to retain the soil moisture.
- Inorganic fertilizers are limited resources because they are made up of elements like potassium and phosphorus that come from mines or saline lakes.

4. Types of inorganic fertilizers

Inorganic fertilizer, also known as chemical fertilizer or synthetic fertilizer, is artificially made in the laboratory and contains all the vital nutrients such as nitrogen (N), phosphors (P) and potassium (K), which are also present in the organic fertilizers. Unlike organic fertilizers, inorganic fertilizers do not have to be broken down before being absorbed by plants because the nutrients present in them are easily absorbed. The purpose of this section is to discuss the various types of inorganic fertilizers with examples.

5. Classification of inorganic fertilizers and their Implications on soils and crops

Based on the principal element(s) the fertilizers contain, inorganic fertilizers are classified into three major types:

- Nitrogenous fertilizers -principal element: **nitrogen**
- Phosphatic/phosphate fertilizers - principal element: **phosphorus**
- Potash fertilizers - principal element:**potash**

Nitrogenous fertilizers

There are 3 types of nitrogenous fertilizer:

- Ammonium/ ammoniac nitrogenous fertilizers
- Nitrate fertilizers
- Amide fertilizers

Ammonium/ ammoniac nitrogenous fertilizers. A property **common to** all ammonium fertilizers is acidicity caused by nitrification with the release of hydrogen (H+) ions. It is therefore, not advisable to use ammonium fertilizers on strongly acid soils except liming is done.

Ammonium sulphate NH_4SO_4	**Appearance:** White and crystalline **Qualities:** Water-solubleMore stable than other nitrogen forms. It is a good source of sulphur on sulphur-deficient soils. **Contains:** 21% nitrogen (N) and 23% sulphur (S) **On application to soil**: The ammonium (NH_4^+) ion nitrifies to release hydrogen ions (H^+) that acidify the soil.
Ammonium chloride NH_4cl	**Appearance:** White, crystalline **Qualities:** Water- soluble. It is a good fertilizer for rice under flooded conditions because the NH_4^+ ion does not denitrify. The chloride is

	also important for the crop **Contains:** 26% nitrogen (N) **On application to soil**:
Ammonium nitrate $NH_4^+NO_3$	**Appearance:** White **Qualities:** Water-solubleIt is less effective on flooded soils due to denitrification losses of nitrogen (N) from the nitrate ion. **Contains:** 34-35% nitrogen (N) **On application to soil**: This fertilizer is hygroscopic (attracts atmospheric moisture). It therefore cakes when exposed to the atmosphere. Leaching losses are higher than the other ammonium forms mentioned above because of the high mobility of the nitrate ion. Ammonium nitrate causes explosion when it comes into contact with oxidizable carbon compounds such as fuel.

Nitrate Fertilizers are hygroscopic; They cake on exposure to the atmosphere. They must be stored in anti-moisture bags. They also denitrify to release oxides of nitrogen and nitrogen gas. They are therefore not recommended for flooded rice or crops grown under anaerobic conditions.

	Calcium nitrate (Ca $(NO_3)_2$. **Appearance:** **Qualities:** Water- soluble - see above **Contains:** 16% N, **On application to soil**:
	Ammonium nitrate (NH_4NO_3) **Appearance:** **Qualities:** Water- soluble - see above **Contains:** 34-35%N **On application to soil**:
	Potassium nitrate KNO_3 **Appearance:** **Qualities:** Water- soluble - see above **Contains:** 13% nitrogen (N) **On application to soil**:
Calcium ammonium nitrate (CAN)	**Appearance:** **Qualities:** Water- soluble - see above **Contains:** 26% nitrogen (N) **On application to soil**:
Ammonium sulphate nitrate	**Appearance:** **Qualities:**

	Contains: 26% nitrogen (N) **On application to soil:**
Amide fertilizers	
Urea $CO_2(NH_2)_2$	**Appearance:** White, granular **Qualities:** It is hygroscopic in its crystalline form. It is the most concentrated solid N fertilizer **Contains:** 46% nitrogen (N) **On application to soil:** Urease, an enzyme, converts urea into ammonium carbonates which release ammonia. When the release of ammonia occurs on (or near) the soil surface, ammonia is lost to the air. If the release occurs near the seeds, the seeds may fail to germinate. Roots may be affected by the toxicity of ammonia.
Calcium cyanamide $CaCN_2$	**Appearance:** **Qualities:** Contains lime and does not make the soil acidic **Contains:** 21-22% nitrogen (N) **On application to soil:** When cyanamide decomposes in the soil it forms ammoniacal nitrogen at a slow rate. The other intermediate products during cyanamide decomposition cause plant toxicity. Calcium cyanamide, which is an expensive source of nitrogen, has to be applied 2-3 weeks before sowing.

6. Management of urea and other nitrogenous fertilizers in agricultural soils

Incorporation of urea into soil will minimize NH_3 losses by increasing the volume of soil to retain NH_3. If soil and other environmental conditions appear favourable for NH_3 volatilization, deep incorporation is preferred.

- Band placement of urea will probably result in soil changes comparable to those produced by application of anhydrous NH_3. Diffusion of urea from banded application can be 2.5 cm within two days after its addition, while appreciable amounts of NH_4^+ can be observed at distances of 3.5 cm from the band. After dilution or dispersion of urea in the band by moisture movement, hydrolysis begins within 3-4 days or less under favourable temperature conditions.

- Point placement of large urea granules will substantially increase the efficiency of urea for rice production. This is the usual practice on small rice production. The usual practice on small rice farms in developing countries is to broadcast small urea granules into the flood water, which frequently results in only 20-30% utilization by the plant. Incorporating the initial dressing of urea into puddled soil can increase efficiency by 35-45%. However, hand placement of large urea particles can double the efficiency to 75-85%.

Puddled soil (poached soil): Soil in which the structure has been destroyed by the physical impact of rain drops, by tillage when wet, or by trampling by animals

Placement of urea with the seed at planting is not recommended because of the toxic effects of free NH_3 on germination seedlings. The harmful effects of urea placed in the seed row can be eliminated or greatly reduced by banding at least 2.5 cm directly below and/or to the side of the seed row of most crops.

Inclusion of acid-forming fertilizers with urea may help reduce soil pH and NH_3 produced during urea hydrolysis. The advantage of acid-forming fertilizers may be reduced in calcareous soils because the $CaCO_3$ may neutralize the H^+ before urea hydrolysis occurs. Volatilization losses of N from surface applications can be reduce if solid urea is applied with solid Calcium salts.

Phosphatic/phosphate fertilizer

Examples:

Single superphosphate	**Appearance:**
$Ca(H_2PO_4)_2.CaSO_4$	**Qualities:**
	Contains: 16-20% phosphorus
	On application to soil:
Triple superphosphate	**Appearance:**
$Ca(H_2PO_4)_2$	**Qualities:**
	Contains: 40-45% phosphorus
	On application to soil:
Diammonium phosphate	**Appearance:**
$(NH_4)_2 HPO_4$	**Qualities:**
	Contains: 46-52% phosphorus
	On application to soil:
Monoammonium phosphate	**Appearance:**
$NH_4H_2PO_4$	**Qualities:**

	Contains: 48-53% phosphorus
	On application to soil:
Rock phosphate:	**Appearance:**
	Qualities:
	Contains: 25-30%
	On application to soil:

7. Management of phosphate fertilizers application

Banding is the recommended application method for phosphate fertilizers:

a) Broadcasting

When water-soluble phosphate fertilizers are broadcast, each granule is surrounded by large volume of soil. In strongly acid soils ($pH < 5.5$) the phosphate granules will be attracted by the alluminium (Al) and iron (Fe) hydroxyl cations to form precipitates and thus reducing the availability of **phosphorus**(P).

b) Banded application

When the fertilizer is banded, there is reduced contact between fertilizer granules and soil particles. As a result, the amount of phosphorus (P) absorbed/fixed is minimal and thus makes it more available to plants.

Potash fertilizers

Potash fertilizers:	
Muriate of potash KCl	**Appearance:**
	Qualities:
	Contains: 48-60% potash
	On application to soil:
Potassium sulphateK_2SO_4	**Appearance:**
	Qualities: 48-50% potash
	Contains:
	On application to soil:
Potassium nitrate KNO_3	**Appearance:**
	Qualities:
	Contains: 44% potash
	On application to soil:

Fertilizer application

In order to get maximum benefit from manures and fertilizers, they should be applied at the right time and in the right manner.

Different soils react differently to fertilizer application. Similarly, the N, P, K requirements of different crops are different and even for a single crop the nutrient requirements are not thesame at different stages of growth. This section deals with the various methods of applying fertilizers to ensure maximum use by crop plants.

Conditions to be considered in fertilizer application

- Availability of nutrients in manures and fertilizers.
- Nutrient requirements of crops at different stages of crop growth.
- Time of application.
- Methods of application, placement of fertilizers.
- Foliar application (application through the leave by spraying liquid fertilizer).
- Crop response to fertilizer application and interaction of N, P, and K.
- Residual effect of manures and fertilizers.
- Crop response to different nutrient carriers.
- Cost of nutrients and economics of fertilizer application.

Note: Fertilizers are applied by different methods for 3 main purposes:

- To make the nutrients easily available to crops,
- To reduce fertilizer losses and
- For ease of application.

Methods of fertilizer application

It is very important to choose the right method of fertilizer application. The method chosen depends mainly on:

- Kind of soil we are preparing
- Type of crop we are planting

- Nature of nutrient we are applying
- Irrigation facility in the area, i.e., whether the land is irrigated or rain fed.

Nutrients to be used by plant must be placed in such a manner that they can be dissolved by the moisture in the soil. The rates and distances that plant food element can move within the soil depend on the chemical nature of the material that furnishes the nutrients and character of soil.

Organic manures are mostly spread uniformly in the field and incorporated several days before planting. The following are the most important methods of application of fertilizers, especially chemical fertilizers. These are discussed below one by one:

Broadcasting - The even and uniform spreading of manure or fertilizers by hand over the entire surface of field while cultivating or after the seed is sown in standing crop is termed as broadcasting. There are two types of broadcasting:

- Broadcasting at planting
- Top dressing - when plants are emerging

Broadcasting manure and fertilizers at planting or sowing time has the following objectives:

- Distributing the fertilizer evenly to incorporate it into part of, or throughout the plough layer and
- Applying larger quantities that can be safely applied at the time of planting/sowing with a seed-cum-fertilizer driller.

Broadcasting at planting is adopted with the following conditions

- When nitrogenous fertilizers like ammonium sulphate, ammonium nitrate and concentrated organic manures, are to be applied to soil that is deficient in nitrogen or where nitrogen is exhausted by previous crops like fodder, and maize.
- When citrate soluble phosphatic fertilizers like basic slag and dia-calcium phosphate, are to be applied to moderately acid to strongly acid soils.
- When potassium fertilizers like muriate of potash and potassium sulphate are to be applied in potash deficient soil.

Top dressing: The spreading or broadcasting of fertilizers in the standing crop (after emergence of crop).

Generally, NO_3 – N fertilizers like sodium nitrate, ammonium nitrate and urea, are top dressed to the closely spaced crops like wheat and paddy rice so as to supply nitrogen in readily available form to the growing plants.

Care must be taken in top dressing so that the fertilizer is not applied when the leaves are wet or it may burn or scorch the leaves. The top dressing of P and K is ordinarily done only on pasture lands which occupy the land for several years.

Side dressing

The term side dressing refers to the fertilizer placed beside the rows of a crop (widely spaced) like maize or cotton.

Placement

The fertilizers are placed in the soil irrespective of the position of seed, seedling or growing plant before or after sowing of the crops. It includes:

Plough sole placement: The fertilizer is placed in a continuous band on the bottom of the furrow during the process of ploughing. Each band is covered as the next furrow is turned.

By this method, fertilizer is placed in moist soil where it can become more available to growing plants during dry seasons. It results in less fixation of P and K than that which occurs normally when fertilizers are broadcast over the entire soil surface.

Deep placement or sub-surface placement: In this method, fertilizers like ammonium sulphate and urea,are placed in the reduction zone as in paddy fields, where it remains in ammonia form and is available to the crop during the active vegetative period.

This approach ensures better distribution in the root zone, and prevents any loss by surface runoff. It is done in different ways, depending upon local cultivation practices such as:

- **Irrigated tracts:** Fertilizer is applied under the plough furrow in the dry soil before flooding the land and making it ready for transplanting.
- **Less water condition:** Fertilizer is broadcasted before puddling which places it deep into the reduction zone.
- **Sub-soil placement:** This refers to the placement of fertilizers in the sub-soil with the help of heavy power machinery. It is done in

humid and sub-humid regions where many sub-soils are strongly acid, due to which the level of available plant nutrients is extremely low. P-tic and K-ssic fertilizers are applied by this method in these regions for better root development.

Localized placement: It refers to the application of fertilizers into the soil close to the seed or plant. It is usually employed when relatively small quantities of fertilizers are to be applied.

It includes the following methods

Contact placement or combined drilling or drill placement: Seed and fertilizer are placed together while sowing. It places the seed and small quantities of fertilizers in the same row. This is found useful in cereal crops, cotton and grasses but not for pulses and legumes. This may affect the germination of the seed, particularly in legumes due to excessive concentration of soluble salts.

Band placement: Fertilizer is placed in bands, which may be continuous or discontinuous to the side of seedling, some distances (about 10-15 cm away from it) and either at level with the seed, above the seed level or below the seed level.

There are two types of band placement

- Hill placement
- Row placement
  - **Hill placement:** When the plants are spaced 90 cm or more on both sides, fertilizers are placed close to the plant in bands on one or both sides of the plants. The length and depth of the band and its distance from plant varies with the crop and the amount of fertilizer as in cotton.
  - **Row placement:** When the seeds or plants are sown close together in a row, the fertilizer is put in continuous band on one or both sides of the row by hand or a seed drill. It is practiced for sugarcane, potato, maize, tobacco, cereals and vegetable crops. Higher rates of fertilizers are possible with row placement than hill placement. For applying small amount of fertilizers, hill placement is usually most effective.

Advantage localized placement

- The roots of the young plant are assured of adequate supply of nutrients,

- It promotes rapid early growth,
- It makes early inter-cultivation possible for better weed control,
- It reduces fixation of P & K.

Disadvantages of localized placement

- Farmers often worry that if crop fails fertilizer is wasted
- Its labour intensive.

Pellet application: In this method, fertilizer (nitrogenous fertilizers) is applied in the form of pellets 2.5–5.0 cm. deep between the rows of paddy crop. Fertilizer is mixed with soil in the ratio of 1:10 and make into dough.

Small pellets of convenient sizes are then made and deposited in the soft mud of paddy fields. It increases the efficiency of nitrogenous fertilizers.

Side dressing

Fertilizers are spread in between the rows or around the plants.

It includes:

- Application of nitrogenous fertilizers in between the rows by hand to broad row crops like maize, sugar cane, tobacco and cereals. It is done to supply additional doses of N to the growing crop. (ii) Application of mixed or straight fertilizer around the base of the fruit trees. It is done once, twice or thrice in a year depending upon the age of the fruit tree.

Fertigation: The fertilizer is dissolved in the irrigation water and applied to the standing crops. It is safe when fertilizers are applied with drip irrigation. Application through sprinkler may cause burning of foliage.

Foliar spray

- Nutrients are applied in the form of dilute solution on standing crop over the leaves of the plants.
- Since there is the direct application of nutrient to the site of metabolism the nutrient use efficiency is increased and quick response is observed by plants.
- This method is more fruitful (convenient, economic and with quick results) when:

- Small quantity of micronutrient is needed to apply.
- It cannot be applied effectively through root or soil,
- There is need to apply partial quantity of nitrogen in the form of urea.

- It is not possible to give the total requirement of major nutrients through foliar application because higher concentration causes leaf scorching, and if frequency of sprays is increased, it increases cost of cultivation.

The minimum safe concentration and frequency of spray solution depends upon the crop, the stage of maturity, season of spraying and the wetting and adhering quality of the spray. In general, 1 to 3 sprays of micronutrients and 3 to 6 of macro element nutrition with wetting the leaves thoroughly in each spray are advised.

8. Which fertilizer should be applied and how?

- Since phosphates generally move only short distance from their points of placement so they must be placed in the zone of root development for better utilization by plant. When phosphates are applied on the surface of the soil after a crop is planted it is outside the zone of root activity and is of little value to row crops in the year of application.
- The placement of water soluble phosphorus in bands tends to reduce contact with the soil and result in lesser fixation than in broadcast application.
- In contrast to phosphorus, the nitrate salts are mobile and move vertically or horizontally within the soil as the water moves. In fine textured soil the movement of nitrogen is restricted.
- Potassium salts are less mobile than the nitrates but are more mobile than potassium.
- In general, N and K carrying fertilizers are more readily soluble than the P material. Therefore, they cannot be safely concentrated in large amounts near the seed or the roots of plants because of the danger of salt damage.
- Likewise, a reduction of soil moisture increases the concentration of the soil solution. Therefore, relatively large amounts of fertilizers placed too near the seed or seedling roots are likely to cause injury during the dry periods, especially when such periods occur soon after the application of fertilizer.

- It is desirable to divide the total requirement of fertilizer nitrogen into several parts to be applied from time to time during growing season. Fertilizers rich in potassium should be placed in a band to the side and below the seed or transplant.

Starter Solution

Solution of fertilizers consisting of N, P, and K is prepared in desirable concentration and directly applied to the roots of young plants at the transplanting time. Such solution is termed as "starter solution". This method allows a direct utilization of cheapest nitrogen and phosphorus sources.

Advantages

- The nutrients reach the plant roots immediately.
- The solution is sufficiently diluted so that it does not inhibit growth.
- Starter solution stimulates growth of young plants.

Starter fertilizer can be used in tomato, peppers, melons, cabbage, cauliflower, and broccoli.

Time of application

Pre-planting	Planting	Standing crops
Organic matter, like farmyard manure, compost and leaf mould are incorporated into the soil well in advance before sowing/ planting of vegetable crops. When this is done these manures get mixed properly in the soil and start rotting to release nutrients when crop plants are in need of them.	**Chemical fertilizers** are applied as a basal dose and also in the form of top dressing. The basal is applied just one day before sowing or planting and mixed or drilled into the soil. Care is taken for the presence of sufficient soil moisture.	**Chemical fertilizers** are used for top dressing of fertilizer, particularly nitrogenous fertilizers is done 15 -21 days after sowing/planting. This is the time when plants get established. **Foliar feeding** of nitrogen and micronutrients is when plants start showing deficiency symptoms.

Precautions in fertilizer use

Fertilizers should be selected on the basis of soil characteristics and by building up a picture of what you know (see visual diagnosis – chapter 2, section 2)

If you know the approximate pH of the soil

- Avoid acid fertilizers in acid soils and basic fertilizers in alkaline soils. This is because the increase the acidity or alkalinity of the soil.
- Micronutrient should be applied wherever necessary. In acidic soils boron and molybdenum, and in alkaline soils, iron, zinc and manganese should be made available.

If the soil has become pale over time

- Improve soil structure through the addition of organic matter and gypsum. Black and alluvial soils should be deep ploughed. Alluvial soils are found on the banks of rivers.

If soil tests are possible: (See soil testing in chapter 2)

- The amount of fertilizer(s) should be calculated based on soil test for balanced use of nutrients.
- Secondary nutrients like sulphur should be used either alone or through sulphur bearing fertilizers. In acid soils, calcium and magnesium should be maintained at the optimum level
- Phosphate rich calcareous soils may show zinc deficiency problems.

Think about the needs of the crop variety

- Use of high yielding varieties, irrigate at an appropriate time and amounts, remove the weeds and get the plant population right. When the precautions mentioned above are taken cares of the continuous application of fertilizers will not reduce soil fertility rather it will help in sustaining higher crop yields.

Chapter 8

Participatory Approach to ISFM Practice

Summary

ISFM is important, cost-effective and has the potential of improving farming efficiency and crop productivity. This is only possible when farmers understand the concept and follow recommended practices in the ISFM approach.

The challenge is how to effectively disseminate ISFM practices to farmers for sustainable crop production.

1 Why are partnerships important in establishing ISFM programmes?

The patchwork of institiutions and individuals that may make up an agricultural development partnerships

Increasingly development initiatives - such as an attempt to increase uptake of ISFM requires partnership working and a participatory approach.

John P Kotter states that for change to happen you need:

- **People to agree on a sense of urgency -** Something must be done!
- **To build a guiding team -** the right people, who can work together and have the ability to make change happen.
- **With a clear vision -** everyone can understand what is needed and why.
- **Can communicate for buy-in -** overcoming confusion and distrust.
- **Can move obstacles** - through creativity or access to power can remove obstructions.
- **Create short term wins -** make progress against clearly identified steps.
- **Don't let up -** keep going till the vision is a reality.
- **Make change stick -** don't let people slide back to old ways.

So it is hard for an individual or a single institution to do this – so partnership skills are important.

2 Why partnerships are necessary?

Partnership working is not only driven out of a necessity to share resources. There is strong evidence that effective partnerships can have greater impact that organisations acting in isolation.

This requires more investment in the setting up the partnership - but the payback should come in the latter as implementation (making the plan work) and mainstreaming (embedding the practice into organizations) should be easier.

The advantages of a partnership approach are:

- **Pooling skills and experience:** Challenges faced by smallholder farmers and other actors in development require diversity in expertise, experience and resources.
- **Increasing capacity to solve problems:** Different views about perceived challenges are more likely to find sound and feasible solutions.

- **Broadening the perspective of the partners:** Build up the whole picture and to appreciate how to act more effectively - generally more diverse groups make better decisions.
- **Increasing the likelihood of implementation:** Partnerships create broader ownership of solutions and increase the readiness to implement agreed plans.

3 The four phases of partnering

When people from different backgrounds decide to work in partnership towards a shared objective they usually need to develop their skills in partnership working, management and communication.

This will involve working differently and trusting others to do things that traditionally the organisations would have done themselves. Partnerships also require a highly facilitative style of leadership - although to some the idea of leadership in a partnerships seems like a contradiction. It is important to remember that most partnerships are informal working groups - and so cannot hold contracts - therefore one partner has to be responsible for holding contracts and ensuring accountability.

Partnerships are increasingly diverse - private sector, public sector and non-government organisations work together. Some partnerships are long-term others are very short-term and task focused.

Partnerships bring together organisations with different internal structures, mandates, objectives and decision-making procedures meet in a common space. However effective partnerships usually share similar values.

Partnerships undergo different stages of development that can be broken down into four phases:

Phase 1: Starting up – exploration and consultation (scoping and identifying):

The length of time that this takes depends on the complexity of the work plan and the pre-existing relationship and trust between the key players in the partnership. It is also helpful if the key players are experienced in partnership working.

The challenge in this phase is to grow a partnership idea together with interested and dedicated people and to do a gap analysis to see which other partners should be involved, or if other changes are needed in the partnership line up – due to conflicting values, priorities or capacity issues amongst the initial group seeking to establish a partnership.

This phase is also about thoroughly testing the potential of the idea and establishing the partnership core group. The quality of trust and relationship building in this phase is a crucial success factor for the subsequent implementation of the partnership's aims.

Key tasks for the start-up phase

- **Explore the prevailing context:** The idea underpinning the partnership needs to be resilient. At this stage it is important to have a clear view of the environment in which the partnership idea should grow: policy, markets, institutions,
- **Test the idea** and players requires working together, learning to keep a process going without forming premature structures. This is likely to include a process of stakeholder analysis (stakeholder mapping) and identification of key partners (and gaps). This requires an exploration of the skills, different interests and motivations and potential areas of conflict between key players. In dialogue the idea will be shaped or the plan to work in partnership will fold.
- **Create a partnership core group**: The key to success of the partnership is a committed core team of different partners. This team functions as the nucleus of the partnership that will carry the enterprise through its difficulties and successes.

The first partnership core group can be small (usually up to 5 people) but it should reflect the diversity of the proposed partnership. The core group in an informal way allows the team members to get to know each other and to build a better understanding of each other's needs, constraints and aspirations. Trust and dialogue makes a partnership work.

- **Grow the idea:** As the idea forms up it important that the core group test the idea informally by talking to experts and potential supporters, integrating feedback and further developing the idea. After a more informal engagement process, more formal partnership meetings are needed in order to secure commitment and create a sense of ownership among partners. Leadership in partnerships is crucial: the commitment and continuity of a partnership core group to lead and coordinate the intervention is essential.

Phase 1 checklist	Who are the actors/key stakeholders that must be involved?
Scoping context:	
Has an assessment been carried out of the current situation and factors relevant to the intervention?	
Reviewing membership:	
Have best practices been researched that could be adopted or learned from?	
Are there clear incentives for people to join the partnership?	
Are the key partners reliable?	
Do the key partners share a common goal and shared values?	
Have potential links to other organisations been identified?	
Reviewing threats:	
Have you assessed potential threats to the success of your intervention?	
Resources:	
Are the resources of different partners known?	
Have funding options for the intervention at this early stage been identified?	
Has the income and livelihood potential of the intervention been assessed?	
Gender	
Has scoping been undertaken to ensure that disadvantaged groups benefit from the initiative?	

Phase 2: Building the partnership (planning, managing and resourcing)

This phase serves to establish the partnership, to give it a formal structure and to agree jointly on goals and planning. Initial structures can be developed, project teams defined and regular meetings planned.

If phase 1 has been undertaken thoroughly and the potential partners have a common understanding of the collaborative approach – phase 2 is likely to be successful.

Stakeholder partnerships are usually formally constituted. With an agreed written document the partnership officially comes to life. This process makes public communication easier. An agreement on short and medium term goals becomes the basis of the implementation plan. Defining the structure of the partnership builds confidence: the contribution and roles of each partner, the allocation of work, all help with the concretisation of the partnership.

In this phase the following three elements need to be taken into account:

Clarify goals and resources: *It* is important to create a climate for shaping the medium and long-term objectives of the partnership and to agree formally the different contributions. This usually requires a workshop setting with key representatives. Complex partnerships may require an external facilitator to guide the discussions.

The workshop should

- Clarify individual objectives and partnership objectives.
- Define how the partnership can strengthen each partner's institutional goals. Establish the financial and contributions are valid. Resources can take many different forms: networks, in-kind contribution, working time, office space, expertise, access to funding, etc. It makes sense to ensure transparent documentation of such a meeting and to consider concerns, suggestions and expectations of different partners. This process can generate a sense of commitment of all the stakeholders to the partnership's goals.

Create the future operational design jointly: The next step concentrates on where to go and what to do next. This too is best done in a joint meeting or workshop. It may require the input of external experts but it can also be a formal planning meeting in which only the partners participate. It always helps to start with a joint analysis of the current circumstances and factors influencing the partnership and assessment of the strengths and weaknesses of the intervention. Results from socio-economic and other research, and market assessment, benchmarking or feasibility studies can be discussed and inform the planning process. This is the point where specific tasks can be allocated to partners or a project team. If not already done, this is the point where intervention activity and task planning becomes urgent: the partnership needs to set out its plan how to achieve economic, social and environmental success.

Consolidate agreements, roles, structures and procedures: With all information from Phase 1 and the joint future operational design, the partnership can now document the planning more formally.

It is time to consolidate a detailed plan of action towards achieving the project goals showing the expected milestones. But it is equally important to agree further joint planning procedures, the internal communication process and rules for external communication. The different roles of partners in the implementation process should be clear at this point.

It is important to ensure that all partners have agreed on the process and the structure of implementation of the identified activities. If not done earlier this is the moment to sign a formal partnership agreement or memorandum of understanding (MOU). There is no single format for partnership agreements or MOU, but there is one single purpose: to move from the more informal Phase 1 to a visible agreement that shows each partner's commitment to the common goals.

Phase 2 checklist	**Who are the actors/ key stakeholders that must be involved?**
Reviewing progress and ownership	
Has Phase 1 been sufficiently completed?	
Do all partners still have a common understanding of the goals?	
Have all partners contributed to the shaping of the goals?	
Are all partners taking ownership of the process and progress?	
Securing a good start	
Will the launch meeting (context, programme, space) secure maximum commitment and ownership?	
Has the launching meeting or a first workshop been designed in a way that partners can together agree the objectives and the implementation plan?	
Documentation and agreements	
Have appropriate forms of agreement for the partnership been identified (argued documents	
Memorandum of Understanding, project plan, press statement, pictures, project	

Is the planning sufficiently documented for all partners?

Roles and responsibilities

Has the allocation of roles and responsibilities been agreed?

Has a communication policy and strategy been agreed?

Is there a realistic time plan?

Have an implementation process and regular follow-up meetings been agreed?

Phase 3: Implementing the partnership (implementing, measuring and reviewing)

The key to a result-orientated implementation of a partnership is to create joint

measures of success; achievable milestones are also important. Partnerships that only stay on the level of planning and dialogue miss this experience of having achieved something together despite all challenges. This is what creates the cohesion a partnership process requires. It supplies the partnership with the resilience also to manage unavoidable difficulties.

Partnerships will have conflicts and disagreements, but if partners have experienced joint success, it is much easier to move forward. They can only do this if they have identified short and long-term benefits at the outset. At this stage, partners need to know that clarifying the goal is not a one-off planning step, but a continuous process. Reinforcing the strength of the vision is crucial for the alignment of partners. Only through getting to appreciate each partner's needs and expectations, being patient and understanding of different points of view, can the partnership grow. The difference in approaches to decision-making and taking action can be accepted more easily if partners understand each others' political and organisational cultures and history. This is the fertile ground for the collective responsibility and leadership a partnership needs. This phase lasts as long as the partnership has defined its implementation plan. In this phase the following three elements need to be taken into account:

Implement and communicate : During implementation it is important to have regular meetings to review the planning, in order to evaluate valuable lessons learnt and adjust strategies. Each of the partners does not need to participate at every stage, but all partners should be part of crucial planning meetings. It is important to maintain agreed procedures, time planning, and the rules for internal and external communication. If neglected, these can undermine trust and commitment. The essence of partnerships is quality communication and relationship management also in the implementation phase.

Build prototypes and celebrate success: Partnerships often go through learning experiences, it is important to generate small success stories that can function as so called "prototypes" of what the partnership is likely to achieve. The partnership can learn from them and proceed to further implementation. Focusing the attention on areas where success can be harvested more easily does not mean losing sight of the larger goals. It follows the principle of iterative learning: where success and failure, mistakes and achievements are jointly evaluated, the likelihood of achieving the objectives set out at the beginning increases. It also helps with understanding where partners need more capacity building and can promote the active search for opportunities to broaden their knowledge and skills. It helps to have a ritual of celebrating success among partners, and combine this with external communication, public events or publications.

Establish feedback mechanisms: Partnerships require project management skills and tools. They also need to create monitoring and evaluation systems that help the key actors stay on track and learn from experience and feedback (like establishing an on-farm farmers monitoring, evaluation and feedback committee). The challenge in partnerships is to develop feedback mechanisms and monitoring systems that are jointly designed and agreed upon. These can be adopted from existing project planning instruments, but adjusted to the specific needs of the partnership, taking into account the differences of the various institutions: private sector, development cooperation, public sector, civil society and community based organisations all have their own type of monitoring systems. The key to outcome and impact monitoring in partnerships lies in the willingness of partners to regularly discuss and lessons learnt and their ability to learn quickly from successes and mistakes.

Phase 3 checklist	**Who are the actors/ key stakeholders that must be involved?**
Implementing, measuring and reviewing	
Is the implementation plan in place and do partners engage with it regularly?	
Are there identified areas where you can have success stories more easily?	

Have indicators for good internal communication been agreed?

Are there agreed rules for external communication?

Have measures to foster relationships and trust been designed together?

Is it clear what further support is needed and who needs to know about progress?

Have performance indicators for monitoring progress been identified?

Have internal capacity building measures been put in place – if required?

Phase 4: Sustaining, replicating and scaling up (revising, institutionalising and moving on)

If implementation has been successful and goals have been achieved, the question that arises is whether the partnership has fulfilled its purpose and can be dissolved, or whether a further development of the partnership's achievements through scaling-up, replication or institutionalization is now desirable and desired. Some partnerships may need to constitute sustainable structures. If partners agree to take the partnership to the next level, Phase 4 needs to concentrate on building appropriate organisational and management structures without losing sight of the crucial role of people and process. It is also about creating the next level core team by inviting and integrating new participants. This often requires building more formal structures and developing partnership governance systems. Again, leadership is crucial here: The commitment and continuity of a new core team is essential in taking the partnership to the next level. This phase has no time limit: it depends on the new planning arrangements. In this phase the following three elements need to be taken into account:

Build new core team: An important factor in the transition from a partnership to a more institutionalised structure, to a replicating or to a scaled-up project is that in principle the future set-up needs to stay faithful to the key objectives and values of the initial partnership. It is important to ensure that the major aspects of the partnership - transparency, multi-stakeholder representation, dialogue, ownership, consensus-building and outcome orientation - are also reflected in the scaling-up, institutionalisation or replication.

Although the new structure will develop its own dynamic and build a new identity, its origin needs to be recognisable, whether this is to be reflected in the governance structure, the openness towards iterative learning or the willingness to partner across sectors. The transition from the initial partnership to the next level requires similar engagement processes to Phase 1: a new core group of committed people needs to be build.

Create new structures: To move from a small partnership or a more loosely structured initiative to an institution, a scaled-up structure or into replication is not necessarily an easy process. Roles change and decision-making structures have to become more efficient. Often the partnership activities need external support and advice in how to do this best. Management structures need to be put in place and new people need to be brought on board. With this, informal relationships become more formal and the internal environment of the next level partnership or initiative changes. New support and funding strategies need to be developed. This also can require getting some form of political support.

Build learning and steering structures: With scaling-up, replication or institutionalisation new challenges emerge. Potential beneficiaries, partners and funders may want to know more about the impact of the next level partnership and will ask for more elaborated monitoring and evaluation instruments. If local or international funding is involved, there will be a request for independent external evaluations. The visibility of the partnership will grow and so will its exposure to external interest and criticism. New partners may want to join and the initiative needs to find agreed mechanisms of how to integrate a growing number of interested stakeholders. If the next level partnership remains a truly cross-sector partnership, it needs to develop governance mechanisms. At this point, partnerships often develop into institutions that can be membership based or independent but governed by a council or board of relevant cross-sector stakeholders and partners.

Phase 4 checklist	Who are the actors/ key stakeholders that must be involved?
Revising, institutionalising and moving on	
Has the scale-up and replication potential of our partnership been adequately assessed?	
Are there partners or other collaborating actors who could become promoters for scaling-up?	
Is it clear what the appropriate form of structure or institutionalisation is?	

Are relationships being managed so as to maintain engagement, and ensure trust, ownership and commitment?

Lessons and next steps

Have new stakeholders and partners who can take the partnership to the next level been identified?

Have the strategies been re-evaluated and adjusted for Phase 4?

Have the lessons learnt been evaluated and integrated in the next level of planning designs?

Has the need for a governance system that the next level may require been assessed?

Is there monitoring and evaluation system in place capable of supporting learning and review?

4 The Participatory Learning and Action (PLA) approach

The participatory learning and action approach (PLA) focuses on dissemination and adaptation of recommended ISFM practices, and has been found very useful to researchers, extensionists and farmers.

What is participation and why is it important? Participation, as used in this unit, implies active partnership between policy makers, planners and officials and beneficiaries (stakeholders). In this partnership participants are organised into action-oriented groups that eventually take responsibility for their own development. There is active interaction among all the partners as they engage in joint analysis, drawing of action plans and the formation of new local institutions or strengthening of existing ones. The active involvement of all partners in a program has the following advantages:

- There is understanding, consensus building and wise decision.
- The program leader has better information about people's wishes and needs.
- The disadvantaged in society have the opportunity to take part in decision-making.
- Power structure are changed: Decision-making process is bottom-up instead of top-down.
- It makes programs legitimate, speeds up change process and reduces resistance to change.
- It ensures effective resource mobilization, creativity, self-reliance and encourages initiative.

a) Participatory learning and action (PLA)

Participatory learning and action involves a team of change agents working with farmers and other stakeholders including soil fertility experts and input-dealers in a 4–phase process comprising:

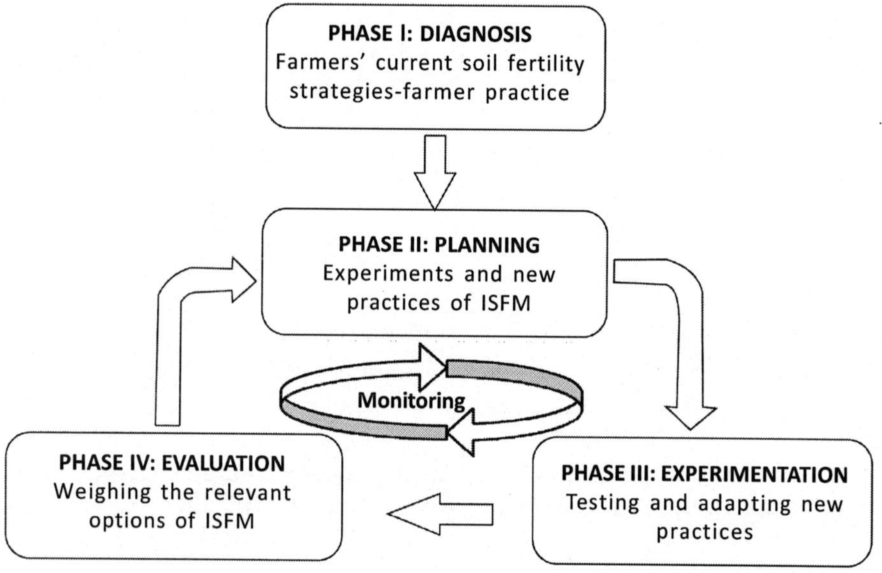

Participatory Approach to ISFM Approach (Adapted from FAO, 2000)

The process essentially seeks to find ways of improving farmers' capacity to analyse soil fertility and find alternative ways of managing it for sustainable crop production.

Phase 1: Diagnostic

The diagnostic phase of the participatory approach to ISFM involves a number of processes with the aim of providing an understanding of farmers' current soil fertility situation and management strategies, and arousing their interest to seek for improvement.

Audit

The usefulness of the current strategies is assessed taking into consideration diverse farming systems and farmers' circumstances. Also information and communication networks and stakeholders are identified including: farmer-based organizations, input-dealers, bankers and policy-makers that can play an active role in the dissemination of ISFM technologies are identified.

5 Participatory rural appraisal (PRA)

This employs participatory rural appraisal methods including community-level meetings, mapping of natural resources, transect walk, analysing soil fertility management strategies, classifying farms, selecting and grouping farmers and drawing of resource flow maps.

Participatory rural appraisal (PRA) is an inclusive, systematic, structured activity that is carried out in the field by a multidisciplinary team to gather information quickly. It is used to effectively diagnose community problems with the active participation of the people in the community.

The use of participatory rural appraisal help to create for the discussion of gender-based ISFM objectives and methodologies, and the identification of land units, soil types, land uses, areas of high and low soil fertility, causes and possible ISFM practices.

Other participatory rural appraisal methods or tools are trend analysis, seasonal calendars, ranking tools (pairwise and matrix) and historical profiles or time lines. At the end of the diagnostic phase it is expected that the selected farmer learning groups should have the same level of information and ready to plan actions.

Advantages of participatory rural appraisal

Inclusion

- Marginalized communities and groups are encouraged to analyse local conditions, prioritise them, and take action. This is a sure way of empowering the marginalised communities.
- It helps to pursue community-based processes for development, including appraisal, planning, implementation, monitoring and evaluation.

Mutual respect

- Scientists use it to identify local priorities for research and initiate participatory research. The scientists become more receptive to local knowledge and recognise that farmers are able to design, conduct and evaluate their own experiments.

Culture change

- It ensures a reorientation of government, researchers, NGO workers and trainers towards the culture of open learning thus moving away from top-down procedures.

Participatory diagnosis: The use of participatory rural appraisal tools enables us to carry out a participatory diagnosis (PD) process in the community. Participatory diagnosis is more than a process to extract information from farmers. It is the first step in engaging with a village as partners in finding solutions. A participatory diagnosis process should:

- Identify and prioritize which problems to solve or opportunities to develop.
- Identify who in the village is most affected by these problems, and
- Nominate who in the village will be responsible for working together to solve these problems.

The outcome of a participatory diagnosis process is an agreement on which problems to solve or opportunities to develop, and how to work together to find solutions.

Phase 2: Planning

After the diagnostic stage, there should be a time lapse before the planning phase to enable farmers discuss the findings and reflect on how they could improve their soil fertility strategies.

The greatest challenge of this stage is the sustenance of farmers' interest and participation. Once the interest for new and/or alternative ISFM strategies is established, meetings with the farmers and other stakeholders should be organized to discuss how to start with a dissemination and adaptation process for ISFM strategies. The action planning phase deals with determining what kind of ISFM strategies will be adopted by farmers and by which farmers.

In the planning phase, farmers' solutions to the constraints and problems, gaps in knowledge and training needs identified in the diagnostic phase, as well as the alternative strategies with their practical and organisational implications for implementation are discussed. This is followed by drawing planning matrix (action plans) of the new strategies the farmers plan to undertake, proposed resource flows, timelines and the expected improvements. The farmers then discuss the technical and organisational implications of activities, agree on how to organise inputs and sets dates for demonstration and training sections. At this stage, the financial implications for implementation of the proposed strategies should be discussed thoroughly with all the stakeholders concerned for commitment and support for implementation.

Phase 3: Experimentation

The experimental stage where action plans that have been decided upon in the action planning phase is carefully carried and monitored by the team of change agents and the different stakeholders.

The experimental phase is basically fieldwork and involves designing and setting up of ISFM technology field experiments, monitoring of experiments, field data collection and analysis, and discussion of results with farmers.

Key activities of the experimental phase of a participatory learning and action exercise:

- **Meet with farmers to finalise experimental design:** Review the planning matrix and discuss trial design and monitoring issues. Fix procedures for layout and monitoring experiments, and set date and venue for the demonstration of the designed experiments.
- **Layout experiment:** Train participants on how to set up an experiment to enable them set up the experiments independently on their farms. The experiment should be adapted to the prevailing conditions and therefore the training section should include site selection, experimental plot size, border effects, marking out plots and carrying out treatments. The training should begin with explanation of the principle behind the experiment, followed by layout of specific treatments and how, when and what to monitor.
- **Monitor experiment:** Visit the experiments regularly to collect data on:
    - Performance of the ISFM practice or technology (farmers indicators);
    - Factors that may affect how ISFM technology performs: environmental factors and management practices;
    - Socio economic factors that might influence the adoption of ISFM techniques by farmers; and
    - Farmers' perception about the ISFM technology.Literate farmers should be encouraged to keep records as much as possible and illiterate ones to use forms and symbols to do it. The field visits will allow dissemination of information among the famers and to others in the community.
- **Manage data:** After collecting the data, it should be analysed and presented in a user-friendly manner. The data should be managed

in a way that will facilitate communication between farmers and the team. Calculate average values for all the performance indicators identified by the farmers. Technological, economic and social analysis could also be done to generate more specific information. Further, adaptability and cost benefit analysis can be done. The results should then be drawn up into series of visual aids for discussion.

- **Discuss results with farmers:** Begin the presentation with a review of farmers' indicators for success. Follow this by the presentation of the averages calculated for each indicator. Then, open up discussions to examine the results and the differences to get farmers' views on the results. End the meeting with farmers' suggestions on how follow up on the ISFM implementation. Note that farmers may decide to reject the new practice or try it out the second season with an adapted design or proposed widespread trials.

Phase 4: Evaluation

The evaluation phase aims to measure the progress being made in the participatory extension cycle for ISFM. This requires a setting up of a participatory monitoring and evaluation system that enables the different stakeholders to analyse specific activities.

The monitoring process starts with the planning and continue through the cycle to ensure that planned activities are carried out when due.

The final assessment, which is the evaluation, is done over a few days at the end of the cropping season.

Three key activities of the evaluation phase

- **Conduct an introductory community meeting:** Discuss objectives, methods, procedures and timetable for the individual trial farm evaluations.

- **Farmers to evaluate their trial farms:** Using the planning map, the farmer group discusses the experiment – major activities (e.g. field and crop rotations, erosion and its control, livestock management and its components, resources entering the field, livestock and other farm components). If everything went according to the plan (i.e. plan unchanged), then it becomes map of implemented activities. Any activity that was not carried out is cancelled.

- **Conclude meetings for each farm group and whole community**: After the evaluation, a concluding community meeting is held where each class presents its reports for a discussion of the findings. Lastly, timescale for starting new planning phase is then set.

From the discussions, it should be clear that change agents have an important role to play in the promotion and institutionalization of participatory learning approaches into ISFM extension.

Dissemination of ISFM information by change agents among farmers and other stakeholders in Ghana is important, but it should be done with a good understanding of sound ISFM practices and participatory learning processes for sustainable crop production.

Chapter 9

Economic (Profitability) Considerations of Integrated Soil Fertility Management

Summary

Food production starts with a good soil. No matter the kind of seed and good agronomic practices that are put in place without a good soil yields will be low and hence income. ISFM is a technology that helps to solve the problem of soil infertility and production.

Farmers only accept and adopt a technology if they see it to be superior to what they have been doing and it makes business sense. Farmers will not worry about the bigger picture of food security or increased production – they need to make a profit to support their household.

Chapter 9 shows what the farmer will go through to determine the viability of the technology and hence be willing to accept it. It includes the validation of the technology in terms of input, output, income and benefits. The benefit-cost ratio based on gross margins is used to decide whether the farmer will adopt or not.

Evidence from the field so far shows that the technology is worth adoption if it yields up to 50% more than the conventional methods. But the cost of borrowing money in Africa is high (money is a farm input) so when money is borrowed to introduce a new technology, the returns on investment needs to be high enough to cover the interest on credit.

1 Introduction

Farming is a business. It is the income source of the farmer and the source of livelihood and as such represents the commercial enterprise that engages the largest number of people in Sub-Saharan Africa. According to the World Bank Factsheet agriculture employs 65 percent of Africa's labour force and accounts for 32 percent of gross domestic product. This makes it a very important sector of the economy.

Farmers spend time, knowledge, expertise and money on the business of farming. They could have sold these resources for income to take care of their families but they decided to use them for the purpose of agricultural and livestock production.

Like all for-profit businesses the farmers aim should be the highest possible profit.

For smallholder farmers however they also need to consider the level of risk they are able to absorb. For most smallholder farmers feeding the family will also be a consideration. This relationship between risk and reward is critical to understanding how farmers will react to emerging technologies.

To get profit the farmer needs to plan the business at least for the production season. This is very important to help calculate the needs for the farming season. This will include the amount of labour needed, the volume of seed to purchase, the number of bags of the various fertilizers and possibly the need for loans to help to finance the purchase of inputs.

Capital (money) is an essential input in ISFM where new (improved) seeds may need to be purchased each year and fertilizer and other soil improvers are added periodically. Money has a cost and by international standards the cost of borrowing in Africa is very high.

Factors that assist uptake of ISFM

- **Access to inputs** – Store or other services that get seeds, fertilizer and inoculants to the farm gate in a timely and cost effective way – preferably supplied by knowledgeable and committed teams.
- **Affordable product pack sizes** that resource poor farmers can access – this may be 1kg fertilizer packs and small seed packages and selective herbicides – and they need to be the right seeds and fertilizers to meet current local needs.
- **Access to finance** – Savings or credit.
- **Secure land tenure**- To make it tenable to make a long-term investment in the soil.
- **Access to labour** – Especially where the recommended approaches need more labour to be deployed at key times.

Factors that assist commercialization of farming

- **Access to information** – Timely and accurate information on market prices.

- **Access to inputs -** for ISFM that's seed, fertilizer (chemical and organic) possibly herbicides/ weedicides and pesticides.
- **A regulatory environment that ensure that farm inputs are of a high quality -** devoid of poor quality inputs and counterfeit products. Is there a need to improve regulations and or enforcement of existing regulations on quality, grades, and standards of agricultural inputs?
- **Strong unmet or growing demand for the crop.**
- **A road or other suitable transport infrastructure -** to get goods to market/ processing plants.
- **Organisations in place that can facilitate the sale of the crop to markets outside the immediate locality -** paying a premium for high quality outputs and stopping dishonest dealing by brokers.
- **Organisations in place that can facilitate a time shift in the sale of crops -** for crops where the price rises post-harvest then warehouses and/or warehouse receipt systems help farmer to farmers to maximize their return.

2 Making a Plan

To plan the farmer needs to:

2.1 Set a target (sometimes called a goal or objective)

Targets have often been expressed in terms of maximizing yield with the assumption this will always lead to maximized profit. However, maximizing yield and profit could lead to different crop choices and different inputs being selected and applied. Where food security is the main issue -maximizing productivity is important. Where expensive inputs have been purchased profit is very important to pay back the investment and re-invest in the future.

Farmers objectives in any given part of Sub-Saharan Africa there will be a dominant faming system. The nature of the dominant system has a big impact on the farmer's objectives. This in turn impacts on the way ISFM will be approached.

Within each farming system, however, we will likely find much variation:

- **More wealthy farmers** will usually have larger farms and will produce a surplus of crop products that can be sold in the market - probably locally. They are more likely to feed crop residues to their own livestock and may even purchase crop residues from surrounding smaller farms so that they can keep more livestock. They are more likely to be using improved seed and fertilizer inputs - although not necessarily at the required levels and they can integrate livestock manure into ISFM to improve their soil.

- **Poorer farmers** usually have smaller farms that may not produce enough food to meet subsistence requirements. Crop residues may be sold to larger farms because the farmer cannot afford to purchase livestock. Poorer farmers often lack the cash resources to purchase inputs such as seed and fertilizer. Farming is often only a part of the household economy and family members may be engaged in off-farm income-generating activities or labour. They are probably using saved seeds and struggling to add much organic matter or fertilizer to their fields.

Whilst wealthy and poor farmers may be encountering similar problems in terms of soil fertility management, the entry points for introducing improved ISFM practices and the timescale for farm improvement may be very different.

Farmers do not belong in discrete groups of 'wealthy' and 'poor' farmers but rather we find continuous variation in terms of farmers so-called 'resource endowment'.

If the farmer's objective is food security then the options offered probably have to include a minimum of bought inputs.

2.2 Get information

As a farmer or a potential farmer (new entrant) information is needed to take decisions. The farmer needs to collect historical and projected information about:

Table 1 Information Needs of Farmers for Decision-making

Costs (input including loans)	Procedures (good agricultural practices)	Prices (value of output)
These should be known at the time of planning	Technology Challenges These should be relatively easy to estimate at the time of planting	Markets Storage Often unknown at time of planting and needs judgment to be applied

See section 6b of this chapter for information on how to produce a participatory budget.

3 Steps to Adoption of a Technology

The ethical stance for researchers introducing new technological innovations to smallholder farmers is that they should do no harm. That is the adoption of the technology should not leave the farmers worse off.

There are many examples around the world of new technologies that have created too much supply of a product resulting in a crash in farm gate prices. This can mean that more work has been undertaken to bring in a larger harvest but that at the end of the season farmers are worse off. When farmers have been encouraged to move to growing cash crops the results can be disastrous for resource poor or resource limited families as they may not even be able to eat the remaining produce.

In order to ensure that technologies can work at scale there is a series of steps that smallholders need to follow:

3a Introduction of the technology

New technologies are often brought to smallholder farmers by researchers, extension agents or business people. Collectively these people are known as innovation brokers.

Farmers look for benefits from new technologies especially better yields that could result in higher income. The farmer must be convinced of achieving these improvements to make a trial attractive.

In many cases incentives must go with the introduction of the technology. The innovation broker may have to provide all the needed inputs to create an on-farm pilot. This helps establish the efficacy of the technology but not the economics of establishing it – techniques such as participatory budgeting can help establish this.

Pilots can also happen on research station and shown to farmers on open days.

3b Test the technology

Testing of technology is often done through experimentation with the farmers. Selected farmers are involved in the demonstration at various stages of the process. Details of experimentation are covered in chapter 8 of this book. During the process Farmer Field Fora (FFF) are held to showcase the benefits of the technologies. Questions are answered during the periods.

3c Farmer's evaluate the technology and compare with the existing ones

At the end of the demonstration the participating farmers and those who were involved in the Farmer Field Fora are given the opportunity to evaluate the technology.

The outcome of their evaluation determines whether they will try the technology or not. The demonstration needs to be carried out between 3 to 4 times to confirm the observation of the farmers. Without positive endorsements of farmers the new techniques should not be advocated.

3d Determine the sustainability of the technology (availability of inputs)

When farmers are convinced that the technology is good and have decided to go with it they take a few more things into consideration:

- Will they get all the needed inputs after the demonstrations?
- Will they get them at the right time?
- Can they afford the inputs/the technology?
- Will there be ready market for the produce?

Answers to these questions will make them try the technology on their own which may lead to adoption.

Innovation brokers should work on sensitization of input and output markets as the technology is developed. It is also advisable to ensure that inputs used for demonstrations are purchased through the local network of agro-dealers - so they are getting used to the technology too.

3e Try it on a small scale and evaluate again

Most small-scale farmers will not change completely to the new technology even when it is very promising. Resource poor farmers need to be risk averse. As a result many of them will only devote part of their resources to the new technology while they continue with their normal practice. If they succeed for two or three consecutive seasons then they increase their investment in the new technology. Hence they replace the old technologies gradually.

In adopting Integrated Soil Fertility Management (ISFM) technologies the farmer must be sure that it is superior to their current practice and that the gains are worth the new investments. It is important to think about how a technology could be superior:

- It could increse yield or greater quantity and or quality (grain, stover and seed).
- It could reduce input costs (labour).
- It could reduce risk (early maturing crop that is less likely to be impacted by drought, or be resistant to pests and diseases).

Farmers usually try a new technology and convince themselves about it before it is adopted at scale. In general adoption is gradual - both by individual farmers and within a farming community as indicated by Roger in figure 9.1.

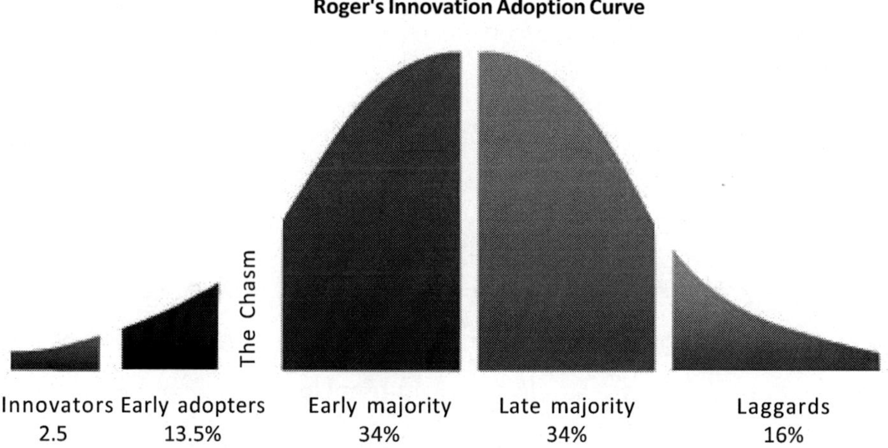

Figure 1: Roger's Innovation Adoption Curve

There is always an element of risk associated with adopting a new technology. So innovators in agriculture tend to be more affluent farmers. This is for two reasons:

- They have access to resources to pay for the inputs
- They are more able to take a risk inherent in a new technology

In some cases other factors inhibit uptake - especially availability of essential inputs

Opportunity costs

Technical superiority is not enough. There will be a range of issues that also need to be considered. These include the things that could have been done instead. In economics these are called the opportunity costs. ISFM looks for organic matter to reapply to the soil. So if the technology requires stover to be ploughed into the field – the farming family must consider the value of the current uses of the stover. These could be livestock feed, building material or fuel. So, the willingness of a farmer to apply the stover to the land is not just a decision about the value of the new technology but the impact it has on the family. If using stover as organic matter in the soil means a huge burden of colleting firewood from the locality it is possible that the labour needed for good agricultural practices will not be available. This is where social science and agronomy need to work together.

4 Key Aspects of the ISFM Approach

Table 2: Some Aspects of ISFM

IS + F	+ M
Improved seeds + Fertilizer	**More ...**
Improving the efficiency of external inputs	Maximizing on-farm recycling of nutrients
	Reducing nutrient losses to the environment
	Replenishing soil nutrient pools

All these involve cost hence the need to assess the extra cost and benefits before adopting the technology

5 Comparing Costs of the Traditional and New Technologies

The positive difference between the traditional technology and the new one must be enticing enough to make a farmer invest in it. The farmer asks himself:

- How much **extra** investment or expenditures do I have to make?
- How much of such will be in capital investments?
- How much **extra** income will I get?
- Is the extra income worth adopting the technology?

Considering ISFM let us find answers to these questions.

5a How much extra investment or expenditures do I have to make?

To establish how much extra investment or expenditure is required. 2 important questions must be asked:

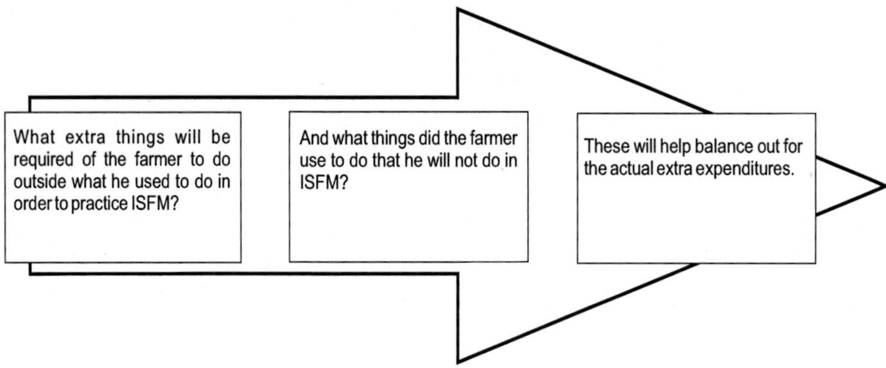

Figure 2: Important Questions to Ask

The answer to the second question is simply nothing. The farmers has to do everything they used to do and in addition they have to do the following new things:

Table 3: ISFM addition to Traditional Methods

Test the soil - ISFM begins with knowing the state of the soil.	This is an activity that most farmers do not do but which is very important for soil management to increase output. It becomes a new activity.	How much does it cost to test soils of a 1 hectare field?
Add extra manure	Normally the farmer uses some manure but not enough to give the desired output. He will now use the recommended amount per hectare.	What is the extra cost of manure?
Add extra or reduce fertiliser.	There is an old general recommended fertiliser	This calls for either extra cost or lower cost.

	rate for maize production that farmers have been using. For ISFM there is the need to test the soil to know what nutrients are low and based on that the required fertilisers and their proportions are recommended.	depending on the results of the test.
Use improved seeds	Farmers often take from their previous harvest as seed to sow but this does not always give the desired yield. ISFM requires that the best seed is used and hence the need to buy at a rate that many farmers may think is too high. Each year the farmer needs to get good seed to sew.	This calls for extra cost. How much is the extra cost or gain?
Use more or less seeds than before	ISFM requires an optimum plant population based on the planting distance and number of seeds per hill. Depending on what the farmer has been doing he may use more or less seeds under ISFM.	Again what is the extra cost or gain?
Use more labour or time	Due to the planting distance and the plant population more time is taking to sow and there is the need to be careful in weeding and performing other agronomic practises hence more labour or time is required.	This again brings extra cost. So how much is this extra cost?
Cover cropping	Farmers do not often do this. If this is adopted under ISFM then there	How much will this cost to do?

	will be an extra cost of seed, sowing and possibly more or rather less labour at weed control.	
Green manuring	This requires seed of the crop to be used, labour for sowing and ploughing cost. The costs involved here are extra costs that will come as a result of adopting this aspect of ISFM and must be taken care of.	Again, how much is this?
Crop rotation	This is a practise that some farmers do hence may not bring extra cost except that there may be an opportunity cost.	You are forced to cultivate a crop you may not have cultivated if not for the ISFM technology.
Good Agronomic Practices (GAPs):	For optimum yields ISFM insists on good agronomic practices which may demand more labour or labour time.	This brings extra cost and must be taken care of in the cost build up. How much will be the extra cost?

The costs involved in doing all these practices are determined and added up. It should be noted that most of the costs mentioned above are extra costs. New costs stand as they are and need no adjustment

Where the ISFM technique replaces something else the saved cost needs to be subtracted from the new ones to assess the real cost of the technology. For example, micro-dosing can reduce the fertilizer or seed use – but may increase the labour costs.

It should be noted that not all farmlands demand all the full components of ISFM. The quality of the soil, the nutrient content and the proposed crop to be cultivated determines the combination of activities that must be chosen and implemented.

5b Example: Pepper growing in Mbanayilli in the Northern Region, Ghana

Table 4 shows the actual operations and costs to the farmer in cultivating one acre of pepper in Mbanayilli in the Northern Region, Ghana.

The left hand columns (2-5) of table 4 show an assessment of the farmers current practice showing how much each of the activities cost.

If the farmer decides to bring in ISFM technologies, there are new operations and other new additional, costs that will be incurred.

For this farmer, the costs are

- Compost preparation valued at GHc 80.00 to meet the need of 1 acre.
- Soil testing which will cost him GHc 70.00.
- Extra labour to apply the compost. The number of person-days of labour he needs will depend on the method of application of the compost (broadcasting or spot application).
- The method of application of compost will have an impact on the required labour too.
- With the application of compost one of the problems will be increase in weeds hence increase in the cost of weed control.
- It is anticipated that the farmer will have higher yield so he will spend more on harvesting - sacks and labour for drying and storage.

If all these costs are captured and put together we have the total cost of production on the ISFM field just as we have on the traditional method.

These need to be reviewed in the context of anticipated gains in productivity and anticipated markets and market prices.

Table 4: Cost of production of pepper per acre in Ghana Cedis (GH₵)

ACTIVITY	Farmers' Current Practice (Mbanayili)				ISFM additional costs			
	UNIT	QUANTITY	UNIT COST (GH¢)	TOTAL COST (GH¢)	UNIT	QUANTITY	UNIT COST (GH¢)	TOTAL COST (GH¢)
INPUTS								
Compost Preparation								**80**
Soil testing								**70**
Seed	Bucket	3	10	**30**	Kg	0.2	**300**	**30**
Land rent	Acre	1	80	**80**	Acre	1	80	**80**
Hoe	Each				Each			
Cutlass	Each				Each			
NURSERY	3 months							
Fencing and shading	Person days	3	7	**21**				
Watering	Days	50	3	**150**				
Watering can	Each	1	8	**8**				
FIELD PREPARATIONS								
Land Clearing	Person days	2	7	**14**				
Compost application	Person days				Person days	2	**5**	**10**
Ridging	Person days	8	7	**56**				
OPERATIONS								
Transplanting	Person days	6	4	**24**				
Fertiliser NPK (2 times)	Bags	1	26	**26**				
Fertilser application	Person days	2	2	**4**				
1st weeding	Person days	4	7	**28**				
Ferttiliser NPK	Bags	0.5	26	**13**				
Fertiliser Ammonia (once)	Bags	0.5	18	**9**				
Fertiliser application	Person days	2	2	**4**				
2nd weeding earthen up	Person days	4	7	**28**				
3rd weeding	Person days	4	7	**28**				
Spraying chemical (Karate)	Litre	1	10	**10**				
Spraying (2 time)	Person days	2	8	**16**				
Water for spraying	Grawa	4	0.5	**2**				
Reshaping	Person days	8	7	**56**				
Harvesting	Person days	100	2	**200**				
Selling								
Store (sacks)	Each	10	1.5	**15**				
Total				**822**				

Source: Yeboah (2010)

Table 5: Comparing Two Systems of Production using Traditional and ISFM Methods

	Traditional method			ISFM Method		
Operation	**Qty**	**Unit price GHc**	**Total GHc**	**Qty**	**Unit price GHc**	**Total GHc**
Soil testing	0	0	0	1	75	75
Soil treatment	0	0	0	0	0	0
Seed	22.5	3	67.5	1	80	80
Land clearing	1	20	20	1	20	20
Ploughing	2.5	60	150	2.5	60	150
Manure application	2.5	30	75	2.5	30	75
Harrowing	0	0	0	0	0	0
Sowing	2.5	20	50	2.5	30	75
Spray pre –emergence	0	0	0	3	10	30
Filling in	2.5	15	37.5	2.5	15	37.5
First weeding	2.5	40	100	2.5	40	100
First fertilizer application	2.5	20	50	2.5	20	50
Pest control	0	0	0	0	0	0
Cover cropping	0	0	0	0	0	0
Second weeding	2.5	35	87.5	2.5	35	87.5
Second fertilizer application	2.5	20	50	2.5	20	50
Monitoring	1	60	60	1	60	60
Harvesting	2.5	20	50	2.5	20	50
De-husking and shelling	1.5	50	75	3	50	150
Winnowing	15	1	15	30	1	30
Drying	15	1	15	30	1	30
Bagging	15	0.5	7.5	30	0.5	15
Transport	15	1	15	30	1	30
Storage (chemical and space)	15	1	15	30	1	30
Marketing (transport, Haulage, tolls)	10	1	10	27	1	27
Sub total		**950**				**1252**
Cost of fertilizer		146				365
Cost of herbicide		0				30
Cost of pesticide		0				0
Cost organic manure		0				375

Sacks	15	30
Hoes	9	9
Land cost	30	30
Expertise	0	30
Extension service		20
Sub total	**200**	**889**
Total production cost	**1150**	**2141**
Yield (tons)	1.5	3
Price per bag of 100kg	90	90
Total revenue	**1350**	**2700**
Profitability	**200**	**559**
Difference		279.5
Increase in Profit		279%

Source: UDS/AGRA Soil Health Project (2014)

6 Comparing of the traditional and new technologies

At the end of the farming season and having captured all the necessary data the following could be compared:

- The costs of the traditional method field (farmer practice) against the cost of the ISFM fields
- The yields of the 2 systems
- The income from the sale of the produce
- The profit from the two systems
- The ease of doing the additional operations
- Ability to raise the additional funds needed to implement the ISFM.

The farmer may only be interested in how much extra expenditure has been made and how much extra income and therefore profit, has been accrued.

If the profit from the ISFM system is higher and appreciable the farmer may decide to go in for ISFM. If the difference is not big enough it is unlikely that the farmer will use the ISFM technology.

As business people farmers need to consider other options of the use of that extra expenditure. The safest investment is said to be the fixed deposit or treasury bills. If the farmer were to put that extra expenditure in the bank for the period would they obtain more than the profit from investing in ISFM? This helps to establish the opportunity cost of the money. There are always conflicting investment opportunities that need to be assessed against each other.

Small-scale rural farmers may see the opportunity to increase their income as a result of the ISFM, but may lack the ability to gather that extra money to invest in ISFM.

It is difficult for rural small-scale farmers to obtain credit for farming hence they may not practise the ISFM technology. In this case the agent introducing the technology must provide a package to assist the farmers begin the process of adoption.

6a Calculating input costs through a participatory budgeting process

Changing farming practice is a big decision for a smallholder - even when they know what they are currently doing isn't working. Any new approach or change in agronomy, including incorporation of ISFM -should only be encouraged on the basis of rational decision-making.

When farmers in Mali were experiencing heavy losses from Striga, they carried out an experiment on 2x25 meter square test plots to decide what to do. With the help of a facilitator the smallholders worked out the inputs required for the new approach - this was combination of more hand weeding and some basic ISFM principles including intercropping.

The first step was for the facilitator to get the local people to explain the inputs they bought and the standard local measures. They bought manure and the unit of measure was a small donkey cartload that cost 500 CFAs.

Labour markets

Some ISFM practices require increased labour during peak periods and/or the hiring of particular services from those having mechanized equipment for the task (ploughing and threshing).

If a farmer has under-utilized family labour and the necessary equipment, this is not a problem. If not, there should be a well-functioning market where temporary workers and needed services can be hired or, as in many countries, a system of community work groups that move from farm to farm. Work on a particular participant's farm is carried out in exchange for work carried out by the participant on other group members' farms (in Ghana termed 'nnoboa').

Many ISFM practices require a significant increase in labour during non-peak periods (e.g. composting, collection and application of manure, building anti-erosion structures, special ploughing techniques to improve moisture retention). Promoters of ISFM often assume that an increase in non-peak labour demands poses few problems for farmers. This is often not the case as farmers often engage in alternative income-earning activities (migration, non-farm employment or self-employment activities) during non-peak periods. This means that in assessing the economic incentives for a farmer to adopt a set of ISFM practices, we need to know the opportunity cost of the farmer's labour.

The opportunity cost of additional labour applied to ISFM is the income foregone from some other activity in which the farmer could have engaged during that period of time.

While the opportunity cost will vary from farmer to farmer, a common benchmark for estimating it would be the average wage for a day labourer in agricultural or non-agricultural activities conducted when it is not the peak season.

When making comparisons – it is the relative cost of the labour that is important – in helping to make decisions. Here the new technology required two hours more labour over a season than the current farmer practice.

Labour analysis	**Current farmer practice**	**New approach**
Land preparation	8 hours	8 hours
Planting	3 hours	4 hours
Weeding + fertilizer	48 hours	50 hours
Harvest	4 hours	5 hours
Post harvest	4 hours	2 hours
Total	67 hours	69 hours

The new approaches require 2 hours additional labour – which can be costed and built into the comparison between the old approach and the new.

If the new approach requires less labour it should only built into the calculation if it is a cost saving – not just a time saving. If the time freed up can be used to earn income elsewhere then it should also be included in the calculation – such approaches might be spraying weedicide versus hand weeding.

Building up a budget for the input costs

The example below is based on an exercise filmed by Agro-Insight – *Let's talk money* is available on the ASHC website or at www.access agriculture.org/node/260/en.

Current farmer practice		**New approach**	
Input			
Millet seed	3,000	Millet & cowpea seed	2,250
Bag	250	Bag	250
Manure	3,000	Manure	3,000
		Fertilizer	500
		Urea	500
		+ 2 days labour	1,000
Total	6250	**Total**	6,475
Return		**Return**	
Millet	**9180**	Millet	3,750
		Cowpea	11,775
		Sub-total	18,850
Profit	**2,930**		**12,375**

In this case working smarter and a very small amount of investment is helping to turnaround the Striga infestation and over a fourfold increase in profits for the farmer.

In other cases the calculations may need to involve factoring the costs of borrowing money to pay for the farm inputs.

Farmers should not be encouraged to use costly fertilizers if they may not be able to market enough farm produce to cover the cost of inputs. Also, as ISFM relies on the cumulative effect of a number of inputs – it is essential that the farmers can see a way to paying each of the inputs suggested.

Implications for extension

Many smallholders will not keep records of yields or input. This means the information they need to make rational decisions is lacking. So, there is a valuable role of facilitating the calculation of the relative costs and making the comparisons.

One of the aims of extension should be the facilitation of better decision-making processes by smallholders. The film *Let's talk* at www.accessagriculture.org/node/260/en/ shows how an experienced facilitator spelt out the different inputs and calculated what each of them cost - and the also the anticipated return.

The example here was based on getting rid of Stiga and using ISFM approaches to improve the soil quality as part of the soil management strategy. So, this is a relatively simple set of calculations.

Knowing that an option could be profitable is the first step - but responsible extensionists also need to think about the markets.

6b Factoring in risk and uncertainty

When working with smallholders it is essential to look at risk and uncertainty and model based on the results of on farm trials. Research station trials are usually hard to replicate on the farm and returns and yields are rarely as high.

It is useful to make an assessment of the major risk factors and rank them according to their influence on farm profitability. Major factors may include:

- Drought during the main cropping season (frequency, magnitude, effect on crop yield)
- Late rains (frequency, effect on crop yields)
- Crop price volatility
- Availability of ready market
- Input prices and availability

If the new approach is making a return using a conservative estimate of the potential harvest - it is likely to be a risk worth taking.

In most cases it is sensible to look at incremental introduction of new approaches. The Striga example above is probably an exception where the control of Striga relies on a concerted effort - not just on one farm, but preferably across a number of adjoining smallholdings

6c Marketing as a risk factor

If smallholders are being encouraged to apply ISFM (or other approaches) to increase yields - it is essential that due consideration is given to modelling likely the impact on future prices. This has been covered at a practice level in section XX on participatory budget setting.

Price is a function of both supply and demand. If supply is increased and there is not additional demand for the goods - the price will fall. Farmers have to work harder to process increased yields - if markets do not exist for their produce they will see prices fall and they may actually be worse off. In the worst cases, where demand does not exist, harvests will be left to rot in the fields and smallholders will not be able to repay loans for farm inputs. The supply of cheaper food may be a good policy aim - but there has to be enough incentive for farmers to keep producing.

Falling food prices may seem a useful policy consideration for non-farming poor families. However, in Northern Ghana a very large number of families have incomes that are related to agricultural value chains. So the relationships between cost and purchasing power are complex.

Factors that need to be present to enable marketing of increased production

- **Markets** - within the region, country or export markets. Is there unmet and growing demand for the crops grown in the target domain? Is there room for making the markets more efficient in a manner that would reduce transaction costs and margins, thereby presenting opportunities to increase producer prices and/or reduce consumer prices?
- **A policy environment** which allows producers to access the markets - negative polices can be tariffs on the movement of agricultural produce, export bans, price controls. What is the role of government (direct participation in purchases and sales/regulatory role in terms of grades and standards, import/export tariffs and subsidy or price support policies)?
- **Transport infrastructure -** allowing timely movement of goods to markets (see below)
- **Intermediaries or other mechanisms to facilitate access to markets** - What is the current structure of the marketing system (many actors and competitive pricing/few actors and unfavourable prices for farmers/an established role for producer organizations in collecting and consolidating production for forward sales)? Are there well-developed value chains that could serve as a lever for the introduction of ISFM practices into a farming system that already has reliable marketing infrastructure?

The economic or business case for ISFM needs assess the package of required inputs and the anticipated value of the increased production and profit that could be generated.

Whilst ensuring that the expansion does not lead to the smallholder overtrading - or running out of cash before the harvest can be used to pay for the input costs.

In deciding on the cropping systems to target with ISFM, attention needs to be given to how well output markets are functioning:

- In addition to working on the output markets work should also be undertaken on input markets. It is important to grow input providers into pilot and trial activity.

Working on input and output markets at the same time as addressing agricultural production is called a value chain approach.

Other Considerations for Decision-making

In addition to the work with smallholders to get ISFM introduced there are other activities around the value chain that may be necessary:

Input markets

- **Input supplier training programmes** that strengthen management skills and encourage the development of distribution networks that link importers, distributors, wholesalers, and retailers efficiently in a way that reduces transactions costs.
- **Encourage effective monitoring of farm input** quality to identify manufacturers and retailers that supply adulterated or fake fertilizer products. This may be through wholesalers or though the Ghanaian regulatory bodies.
- **Encouraging the active participation of a banks and micro-finance agenciesInformation.**
- **Encourage information systems** that provide farmers with periodic (daily, weekly, monthly) localized information on input and output prices so that farmers make more informed decisions. Output markets.
- **Explore more efficient markets** that could reduce transaction costs and margins, thereby presenting opportunities to increase producer prices and/or reduce consumer prices?
- **Explore contract farming models** that link input and output markets, often providing input credit and guaranteed output markets for production.

- **Strengthening of farmer based organizations** so that they can consolidate fertilizer orders for their members and perform some of the procurement and distribution functions themselves.

7 Measures of profitability

Some Measures of ISFM Profitability that the farmer could do. Some are for long-term investments like tree crops.

- Gross Margin
- Farm Income (FI) Measurements
- Benefit-cost ratio (BCR)
- Net Present Value (NPV)
- Internal Rate of Return (IRR)
- Value Cost Ratio (VCR)
- Input-output (I-O) Ratios
- Break-Even Analysis.

For this purpose we will consider 2 of the measures which are commonly used, i.e., the Gross Margin and the Benefit-Cost Ratio.

7a Gross margin analysis

In many farms, the farmer has more than one enterprise he runs and there are some facilities that are shared by the enterprises. These are normally considered as fixed costs. The rate of use per enterprise is normally difficult to compute hence the farmer before declaring profit of the farm goes through the concept of Gross Margins. Gross Margin of an enterprise is the income from that enterprise above the variable cost of that enterprise or the difference between total income and total variable cost of each enterprise. Gross Margin is therefore not profit, it is the enterprise's contribution to fixed costs and profit after the variable costs has been paid.

GM = TVO - TC

GM=Gross Margin

TVO= Total value of output

TC= total cost

To get the profit of the farm the farmer considers the gross margins of all the enterprises and takes off total fixed cost for the period.

$\Pi = \Sigma$ GM–TFC

Π = Profit

Σ GM = Total gross margins

TFC = Total fixed cost

If the farmer has only one enterprise then his Σ GM is the same as his GM and all the TFC is charged against that one enterprise.

7b Benefit-cost ratio

Benefit-cost analysis (BCA) is a technique for evaluating an investment by comparing the economic benefits with the economic costs. The procedure compares costs and benefits over time, ie, if the investment is for more than one year. It helps to determine the potential economic gains/losses of an investment. It could be done before investment using some assumptions said to be ex-ante or at various periods in the implementation (mostly annually) of the investment termed ex-post.

Decision Rule

- If the BCR > 1, investment in ISFM is profitable, but
- If BCR < 1, investment is not profitable.

Why Do BCA of ISFM

- To evaluate the economic merit of the ISFM technology
- To compare the economic merits of different but competing projects eg ISFM technology versus traditional methods.
- To help farmers decide whether or not to adopt ISFM technology
- To ultimately increase producer profits via better planning and efficient resource use.

8 Is it worth doing ISFM?

Having gone through all the discussion above; taken data, analysed and compared the various profit levels the farmer has a decision to make. Should I use ISFM in production or should I go back to my normal (conventional/traditional) practice?

It has been proved with experimental evidence that ISFM is profitable hence scientists are not ashamed to recommend it to farmers.

Example for Study

How to compare costs of N-P-K 15-15-15 versus single nutrient fertilizers, with target yield of 16 tonnes per hectare

Table 6:

Assumptions	US$	Option	Requirement	
50 kg bag N-P-K 15-15-15	25	1 16 x 50 kg bag N-P-K	16 x 25	= **US$ 400**
50 kg bag urea	25	2 5 x 50 kg bags urea 1 x 50 kg bag TSP 3.5 x 50 kg bags KCl	125 + 40 + 140	= **US$ 305**
50 kg bag TSP	40			
50 kg bag KCl	40			

So, in this case using single nutrients would save a significant amount: US$ 400 – 305 = US$ 95 per hectare.

Supporting smallholders to minimize their input costs is very helpful to farmers. However, few smallholders would be applying anything like this level of fertilizer so there is unlikely to be a cost saving – it may however support farmers to get to a more optimized fertilizer application.

Example: Economics of cassava systems

For example, if a farmer who has been obtaining yields of 10 tonnes per hectare of fresh roots with no fertilizer decided to use 16 bags each of 50 kg of N-P-K 15-15-15, the increase in yield required to recover the additional cost of fertilizes (assuming price of N-P-K is USD 25 per 50 kg bag, and price of cassava is USD 40/tonne fresh roots) can be calculated as:

Minimum increase yield required (tonnes/hectare)

$$= \frac{\text{cost of urea fertilizer}}{\text{Price of cassava}}$$

$$= \frac{16x25}{40} = 10$$

As a rule of thumb, a value/cost ratio greater than 2 is needed for an investment to be economically attractive to farmers.

So, in this example, the farmer would need to obtain 20 tonnes per hectare just to recover the cost of the investment in fertilizer.

To take the analysis a step further, the extra costs incurred with use of a new technology can be compared with the additional benefits obtained through the use of the technology.

For example, if in the above example the previous yield was 10 tonnes per hectare but with use of 16 bags of N-P-K fertilizer per hectare the yield increased to 21 tonnes for a cassava variety that responded well to fertilizer ('efficient variety') and 13 tonnes for a variety that responded poorly to fertilizer ('inefficient variety') the economic benefits can be calculated using the following steps:

Step 1: Calculate the change in yield brought about by use of fertilizer

Step 2: How much the change in yield is worth

Step 3: The amount of money required to purchase fertilizer

Step 4: How much money the farmer is left with after deducting cost of fertilizer

Step 5: The value of yield can be compared with the costs, in the value/cost ratio (VCR)

Table 7: Examples of economic benefits of a technology

Step			**'Efficient variety'**	**'Inefficient variety'**
1	Difference in yield	Yield with fertilizer - yield without fertilizer	21 – 10 =11	13 – 10 = 3
2	Value of yield	Difference in yield × price of cassava	11 × 40 = 440	3 × 40 =120
3	Cost of fertilizer	Amount of fertilizer × price of fertilizer	16 × 25 = 400	16 × 25= 400
4	"Profit" or "loss" made	Value of yield - cost of fertilizer	440 – 400 = 40	120 – 400 = -280
5	Value cost ration (VCR)	Value of yield ÷ cost of fertilizer	440 ÷ 400 = 1.1	120 ÷ 400 = 0.3

So, in the above example use of fertilizer would not have been cost-effective; a significantly higher price for cassava would need to be obtained for this level of fertilizer usage to make economic sense.

If additional information, for example on costs of labour, is available, more detailed calculations can be carried out.

Reductions in costs of labour due to herbicide application may prove economical, particularly where persistent weeds are present.

Remember, the best use of a resource should be explored in order to get the best return from the input.

A point to note is that prices of inputs and yields of crops can vary. For example, better incomes could perhaps be obtained by processing raw cassava roots into an added value product. Also less fertilizer should be applied when rains are late and drought is expected, and more fertilizer can be used when rains are on time and adequate.

Chapter 10

Sharing ISFM Information

Summary

Reaching large-scale implementation should be the overriding objective of all projects that have successfully ISFM practices to share provide markets support them. All too often ISFM results are only reported in peer-reviewed journals that are not accessible outside the research community.

Funding agencies often fail to make the production of extension materials an integral and mandatory requirement of research projects.

Unfortunately, in the past, adoption of economically proven technology has often been poor. Some of the reasons include:

- Researchers are not interested or motivated to engage in the extension process.
- Project funding does not include sufficient provision for an extension phase.
- Changes in market conditions (e.g. crop and input prices) make the technology less profitable.
- Changes in emphasis in government policy mean that the technology is either redundant or low priority.

Even with technology proven to provide significant benefits to farmers, it may take five-years or more to achieve any large-scale adoption and refinements may be required to fine-tune the technology to the often changing needs of the target farmers.

1 Development of a communication strategy

A communication strategy should be developed for any ISFM program. In addition to the farmers, an array of different target recipients of extension material should be considered:

- Input suppliers (timing and quantity of fertilizers, seeds and other inputs required; preparation of suitable promotional materials).
- Output markets (amount and timing of production envisaged, prices).
- Credit providers (typical loan requirements per farmer and per hectare; preparation of suitable promotional materials).
- Policy makers (benefits to farmers and the wider community of technology implementation, e.g. expansion of trade, improved food security, increase in production).
- Extension workers (training in technology implementation; knowledge must be greater and deeper than the farmer).
- General public (improved reliability of food supply; reduced requirement for crop area expansion due to improved crop productivity).

These may all require some basic information, but in addition each needs customized information relating to their respective role.

The communication strategy therefore what materials are needed and how they provide the necessary information to each priority group.

Successful campaigns are usually developed in partnership taking on board a different views and experience from different disciples - especially involving research, extension and communications professionals.

Effective communication is an important part of the mix of getting research into use and disseminating ISFM and other development material. This chapter looks at both print and how texts can play a part in spreading information.

The final section sets out guidance for developing an effective dissemination plan for materials that have been developed.

2 Development of extension materials

Extension materials need to be thought about carefully. Spending time in the field and getting to know and understand the target farmers, is a great place to start. This will help to suggest how the extension messages should be directed and what kind of material and media that will be most effective. While leaflets and posters may be sufficient to reach the target farmers, videos and radio broadcasts may be required to inform input and credit suppliers as well as traders. Furthermore, materials may be required to train extension workers.

Irrespective of the media used to communicate with farmers, the material must include the following information:

- A brief description of the farming and cropping system for which the information on ISFM is relevant.
- Benefits to the farmer (e.g. specific improvements to the farmer's cash income, food security or general livelihood).
- Materials and equipment required (quantities of fertilizers, seeds and tools required).
- Procedures (a step-by-step guide to technology implementation including timing and frequency of operations, labour requirements).
- Simple cost–benefit analysis (details of the additional costs and benefits of the technology and an estimate of the overall quantitative benefit to the farmer).
- Risk (susceptibility of the technology to drought, pests and diseases and market failure).

The choice of media can be influenced by a number of factors, such as

- Complexity of the message and terminology used.
- Age and gender of target audience.
- Access to communication devices (TV, radio, mobile phones).
- Literacy (emphasis on visual versus textual material).
- Language (local, english).

An ISFM campaign might include

Written materials

- Leaflets are generally distributed directly to farmers.
- Posters are used to promote technologies in public places or in meetings.
- Manuals are usually used to train extension workers.

Local radio

- Stations can be useful to transmit messages with clearly defined geographic areas.

Mobile phones

- Now used widely in most of sub-Saharan Africa. They provide opportunities to circulate simple text and voice messages.
- Increasingly blue toothed enable phones are being used to share digital content - including films which inform farmers of new agricultural technologies.
- Farmers can also use them to contact agricultural call centres that provide information on crop management as well as weather and prices of inputs and outputs.

TV broadcasts

- Are usually only suitable for messages transmitted over networks with a large geographical spread.

3 Creating smallholder-friendly printed material

From time to time new information materials will be required to help smallholders put research into use. In the past these materials have often not been given enough consideration and as a result have been limited in their success. These simple guidelines are designed to help develop materials that can be used effectively by smallholder.

When producing materials for smallholders it is better to start with a blank page and set the key areas of the technology - and once you have done this identify the messages that will convince smallholders to change.

Consider all of the following criteria:

Technology

- Include realistic suggestions of what can be achieved.
- Include an honest assessment of the impact a technology will make - likely improvements not the best possible outcome.
- Include a clear explanation any risks.
- An understanding of the impact of a technology on a farming family - not just on farm production.

  - but the unintended consequences (stover left in the field and therefore no longer used as cooking.

  - fuel can create a burden of collecting firewood).

Economic data

- A clear cost benefit analysis (even allowing for fluctuating commodity values) helps smallholders.
- Make rational decisions.
- Clear comparisons - e.g. with improved seed and without, with organic material and without.
- Information on likely markets for surpluses - don't recommend investment in scaling up.

Production if there is a risk that the market will be saturated or cannot be reached in a cost efficient way.

Design

- Good design and strong use of colour.
- Layout in a logical order and format.

Text

- Aim for easy to understand information so test your drafts with the target audience to make sure they understand what you are saying
- Farmer's jargon, not scientific language - make sure you know the terms farmers actually use.
- Short words and short sentences make it easier to follow instructions.
- Bear in mind that children of the house may do the reading - where adult literacy is an issue.

Images

- Real photographs - reflecting real conditions - with named farmers in named locations appear to build confidence in the ideas being suggested.
- Ensure that images reinforce all key processes that are recommended - images help lock in meaning for people who struggle to read.
- Clear comparisons - e.g. with improved seed and without, with organic material and without shown in photographs will be very persuasive.

Language

- In the right language – in Ghana ASHC was told 'people who could read, could read English' as well or better than they could read local languages. The partners decided print should be in English and radio and film should be in local languages.

Measurements

- Use of non-conventional methods to explain qualities and distances based on available and familiar items – in addition to conventional methods e.g. plant the maize an arms length (45 cm) apart.
- 50kg fertilizer bags (the bags that the farmers actually used could contain up to 85kg of harvested maize)
- Coke/Fanta crown tops and cutlass blades are always available to farmers. Whilst ensuring that we do not use kitchen equipment for any agricultural practices

Gender

- Ensure that the text and images reflect the differences in the way men and women work. For example in making the ISFM introductory film, ASHC realised that women only usually had access to small animal manure – so more images of sheep and goats were included in the film.

Customisation

- Leave white space for local customisation of the leaflet or poster so that the nearest agro-input dealers and extension service contact details can be written on the print. Instead of saying 'contact your local agro-dealer', leave a space for specific follow-up contacts to be added.

Brands

- Text and photographs within the ASHC materials should not single out brands of agricultural inputs. There was a danger that these brands would be seen as the recommendation – rather than being indicative.
- Chemicals need to be listed in terms of the active ingredients – so that agri-dealers can help farmers to get the right sort of inputs deal with pests and diseases.

- Exception would be where ASHC worked with private sector input manufacturers or suppliers to produce materials or in showing non-standard measurements where branded container may need to be identified to clarify.

Site-specific materials

- The materials developed should be designed and written for a particular locality - and then adapted for use elsewhere.

4 Considerations for producing print material for smallholders with low levels of literacy

When materials are designed for low literacy environments it is important to:

- Create print material that includes related pictures with minimal text. The print material can be used to supplement practical face-to-face training sessions as follow-up or reference resources for farmers.
- Picture-based print materials can only be used to convey simple sequences of actions, thus should not be used as the only form of instruction/training provided.
- Employ a local artist to develop relevant and locally appropriate illustrations for the print materials. Illustrations that are simple and do not provide a lot of detail are more comprehensive.
- Black and white free-hand line drawings with computer enhanced shading are easy to photocopy and thus more likely to be made available to smallholder farmers by local agencies responsible for reproducing and distributing materials.
- Consult with local women and men smallholder farmers and local NGO and government workers when developing illustrations. These local groups can help determine which illustrations are easily interpretable and provide a clear message. At the same time, these groups can provide feedback on the appropriate sequencing of pictures in the instructions. Illustrations and print material should be modified based on feedback to create the clearest picture-based form of instruction possible.
- Test print material in the field with local smallholder populations and local NGO and government staff before producing and distributing print materials on a large scale. Feedback and responses

from testing should be used to finalize the print material to meet the needs of the local providers and end-users.

- Ensure you have the resources and appropriate time-frame to develop effective print materials. In terms of resources, you will need access to local contacts, including farmers, local NGO and government workers, and extension staff to help you produce the material. As well, you will need to reserve time to consult these local groups and test and modify the print materials based on feedback.

5 Producing farmer-friendly SMS and text messages for extension campaigns

This checklist is for the production of down-to-earth printed sms texts for smallholders.

Text messages should be:

Timely – there are two sorts of timely information for smallholders.

- Information that relates to the farming cycle.
- SMS and text messages help to alert farmers to crises or changing information and can give advice on what to do. Dynamic information can include problems like an armyworm outbreak in the locality, short or long-range weather information, market prices of input (seed and fertilizer), or crop output prices.

Advice backed by evidence – preferably farm-level trials rather than just evidence from research stations but including farmer-led innovations and indigenous knowledge as appropriate.

The ethical position for any extension information has to be that it can do no harm and should be capable of offering real help to smallholders. In this context that means not extending the level of risk smallholders are exposed to.

Be actionable – SMS messages need to offer suggestions and solutions – not just offer information (with the possible exception of price). Defining the problem is not enough – wherever possible the farmer should apply their own judgment.

One message – Each SMS should usually concentrate on one key message, unless it is a summary of a number of previous SMSs used to consolidate understanding and present affordable options.

Offer options - Within the ISFM approach there are often options. So a good SMS/ text message campaign will try to set out principles of ISFM and suggest options - without being too prescriptive.

Farmer's should be encouraged to apply their own-judgment.

If the campaign is for resource constrained women farmers - then options that reflect the choices open to them should be prioritized - so very low input options.

Options must be compliant with existing rules and regulations

Be clear - presenting ISFM options has to be done in a way that is clear and easy to understand. This should usually involve a testing stage for all ISFM messages. Quality assurance is an essential part of the campaign to ensure the information is clear and accurate.

Be relevant and not patronizing - smallholders may need some familiar information included in texts to validate the more innovative messages. But sms/ texts messages that include things that are obvious to all smallholders should be avoided. This is a judgment call for the editors but it is essential that the assumed knowledge behind a campaign is tested.

The right language - SMS are by their nature simple messages with a maximum of 160 characters. So short, simple words and short sentences are a must. The right language includes ensuring that farmers' terms (not scientists' ones) are used. There is a further decision that needs to be tested: should the materials be offered in local languages or the main language of the country? In Ghana it was decided that written text should usually be in English. This was because:

- People who could read, read English.
- The written forms of local languages were not precise enough for some of the complexities of the ISFM message; for example distinguishing fertilizer, manure and compost.

If the campaign is being developed for a specific group - e.g. young people - then make sure that you are exploiting the way that young people use sms: this includes emoticons and abbreviations they will understand. Some latitude with conventional punctuation is probably acceptable - if the meaning is clear.

Spoken word generated dial-out phone messages can be easily developed in local languages where this technology is an option.

You only have 160 characters (in the case of Tweets 140), so make sure they all work really hard. Remember a space is a character.

Be realistic - are the ISFM technologies on offer possible for smallholder farmers to replicate, given their limited resources and the other challenges they face. This would include:

- Ensuring units of measurement are presented in ways that can be replicated with a minimum of equipment.
- Ensuring that inputs are described in a way that is clear to farmers.
- Ensuring that inputs are available; for example single nutrient fertilizers may be the best way to address soil fertility, but if agro dealers only stock blended fertilizers it doesn't help.

Localized information - depending on how the database of farmers' sms details has been built depends on how customized materials can be; for example options of crop varieties can be pre-selected based on what works in that locality (and what is available).

It is important to know the level of localization that can be used to select recipients for different messages for example to link smallholders to input dealers with a particular variety of improved seed. This is easy to do as a response to a query - but may not be possible as part of a text out campaign.

Customized crop specific information - again depending on how the farmers' details have been collected, smallholders should be able to select SMS messages relating to crops that are important to them.

Feedback options - where possible feedback options should be developed. This could include links to input suppliers/prices or market prices in different localities. It could include a plant doctor type of service.

Signpost to other sources of information - using short URL codes can mean that website for further information can be included in an SMS.

Manage expectations - plan the text campaign. Try to make smallholders aware that further messages are going to be coming in relation to this subject i.e. +4, (i.e 1 of 5 messages in a set)+3, +2, +1, 0 showing the countdown at the end of the message.

Overall the campaign needs to solve real problems or present real opportunities to smallholders. The information offered also needs to be regularly updated.

Maize mono-crop in Ghana (draft SMS campaign for Northern Ghana)

This 20 xsms/text messages campaign has different options for one of the messages depending on whether it is the minor season or major season. The messages utilize most of the available 160 characters to try to share clear messages about what smallholders should be doing. These messages were developed for a specific area and will need substantial adaptation for other areas. ASHC can help customize your messages to the needs of different smallholders, crops and locations.

1	Follow these 20 SMS tips & grow more maize! You need light well-drained soil, at least as deep your foot (30cm). shallow soil? Try upland rice, cowpea & cassava!	160 characters
2	Clear dense vegetation with herbicide/slash. Burn only branches thicker than a finger. Burning kills helpful organisms & makes soil blow/wash away – add manure.	159 characters
3	Maize needs lots of water-raised earth bunds, trenches or applying mulch (slashed vegetation) keep rainwater in soil - when water runs-off so do soil nutrients.	160 characters
4a	Minor season–choose a fast maturing, drought-resistant variety. Mamaba's high yielding, quality protein maize - short stover. Obatampa's better in major season.	160 characters
4b	Major season - choose a high yielding quality protein maize with long stover such as Obatampa. Mamaba's maize seed is better in the minor season.	145 characters
5	Maize planted in rows is easier to weed & apply fertilizer. Make rows 2 cutlass blades apart & planting holes 1 cutlass blade apart in row. Plant 2 seeds/hole.	158 characters
6	Seed test: arrange 100 seeds on a wet towel on a plate add water daily. If less than 60 germinate after 10 days increase seeds/hole to 4 – but thin to 2 plants.	159 characters
7	2 plants per hole is ideal-Either plant 3 seeds/hole & thin to 2.OR plant 2 seeds per hole & gap fill soon after the plants emerge. 3 seeds/hole = skinny maize.	159 characters
8	Micro-dosing: small amounts of NPK fertilizer applied at planting time ensures NPK is in place as maize seed needs it to grow. Fertilizer mustn't touch the seed.	160 characters
9	NPK fertilizer improves a maize harvest. At planting, apply 50Kg (GH40) get 10 bags/acre, or 100Kg (GH80) get 25-28 bags-Fertilizer mustn't touch the seed.	154 characters

10	When preparing the plot for planting – make 2 holes – a thumb length apart. 2/3 maize seeds in 1 hole & 1 heaped Fanta top of NPK in other.	139 characters
11	2 weeks after planting weed using a cutlass. Do not use herbicide at this stage as it may damage the young plants. Also thin the seedlings to 2 plants per hole!	160 characters
12	If you didn't microdose at planting – top-dress a heaped Fanta top of NPK/plant a thumbs length (5cm) from the maize plant.	159 characters
13	Poultry manure can be used instead 6 weeks after planting remove the weeds by hand or use a herbicide (Round-up) - requires 1 litre/acre. Weeds compete with the maize for nutrients in the soil.	157 characters
14	Option 1:after 6 weeks -make a small hole a thumbs length from the maize plants. Add 1 heaped Fanta top of urea to hole & cover immediately (Option 2 follows).	160 characters
15	Option 2: after 6 weeks apply a level Fanta bottle top of ammonia as a top dressing.	84 characters
16	Harvest when maize as soon as it's matured (look for the black eye). Late harvest means termites, rodents & birds take 3 cobs in 10!	135 characters
17	Husk maize cobs in the field to avoid transporting weevils from the field to the cribs. The husks will improve the organic matter in the soil.	141 characters
18	Dry maize in cribs for aeration & faster drying. Crib should be an arms length wide, as tall as you & as long as necessary. Maize dries better in a narrow crib.	160 character
19	Don't spray insecticide of maize for storage should rather use superactellic during storage – the insecticide will not touch the maize & is safer.	146 characters
20	Return as much organic matter as possible back to the field: manure, crop residues, weeds, compost all help. Fertilizer works better when organic matter is added.	160 characters

6 Producing a dissemination plan to scale-up ISFM approaches

A dissemination plan makes sure everyone knows what is being proposed and what will happen. It ensures that the messages developed get to the key audiences in an efficient and effective manner.

It sets clear targets for the campaign and it maps out the resources (people and things) needed to make the campaign a success.

Step 1: Is the vision clear?

You need to be able to clearly communicate the vision for the campaign it two sentences. If you can't do it now you must be able to do it at the end of these 12 tasks.

Your vision may be very focused, about encouraging farmers to use a new variety of seed - or it may be broader about encouraging farmers to use a range of practices.

If you can't easily explain what you are trying to do - you will not be able to bring others with you.

Step 2: What are your constraints or targets you have?

You need to know what budget you have and any other constraints or targets you face. This can help drive you to be more creative - but you can't plan effectively without knowing the budget and any other targets. For example any funders will want you to target women farmers. Agriculture seasons may also present you with an unmovable timescale.

During various interactions with ISFM partners, including the write-shops, you will identify a number of stakeholders who are willing to participate in the dissemination campaign at different levels and in various capacities. Some partners e.g. the Ministry of Agriculture, NGOs involved in extension, media houses, information service providers, may work closely with research institutions to ensure dissemination of the ISFM information materials to the farmers and other target groups, while others such as some private sector players could potentially support the campaign activities through provision of resources such as funds, transport, etc. Think about what is required from these stakeholders in order to have a coordinated and effective campaign?

Step 3: What is/are the target group(s)?

Often farmers are those that will make key changes in practices on their farms - but for them to make those changes others need to change as well. For example, if the campaign involves a new variety of seed or fertilizer then the agro dealers may need to know about how it should be managed or sold. Extension staff, whether they are from the public extension service, an NGO or a farmer cooperative may need to know how to advise farmers. So thought needs to be given on who the scale-up campaign needs to engage with - some will be target groups others will be partners - see step 7.

Make a list of the target audiences in as much detail as you can - or how you plan to get information to the targets. E.g. targeting farming families through school projects. Remember if you say everyone - you will probably reach no one!

Step 4: What changes in attitude, knowledge, behavior and practices are sought as a result of the campaign?

Usually at the end result we are looking for in an ISFM dissemination/scale-up campaign is for farmers to be doing something differently.

Finding ways to produce organic matter that can be combined with fertilizer, making decisions about the best way to use the small amount of fertilizer that they can afford. Perhaps we want extension staff and agro dealers to be advising farmers in a different way to support the changes looked for at farm level. Also think about the long-term impact that the campaign could bring about.

For each of the audiences set out in task 3 (above) you need to state what change you are trying to affect.

Step 5: What are the key benefits for each target group?

Before you try to set out the messages it is essential to map out the benefits for the target audience and find creative ways to express these. For example doubles your harvest for the cost of one bag of maize -(or even makes you more money/ food secure which is really the benefit)

Try to avoid listing too many features - or background information. Clarity is needed.

Step 6: What are the key messages for each different group.

They must raise awareness of the benefits of a practice or technology or of combining practices and technologies. But you also give people information they need to make decision. You also need to ensure you have included the information needed to implement a new practice or technology.

Step 7: What language do you plan to use? How you will test the translations or text

Language is the most important aspect of communication. Most countries in Africa have many languages but there is always a common language that most people are able to speak and often read or write.

When communicating to farmers, it will be cost-effective to use a common language spoken across regions, as opposed to using several local languages. For each of the target groups, think about the most suitable/ effective language for communicating with them.

Consider the communication media that will be used with this language. For example is it practical to develop video, audio or print materials in the local language? Does the language have symbols and syntax that will clearly communicate the messages? Are there some audiences that need materials in specific languages?

Will there be need for translations? How will you test these are correctly portraying the messages you need to deliver?

Step 8: What are the communication channels and tools you plan to use?

Communication channels and tools that are feasible will be a compromise between what is desirable and what is practical given the available resources. How will information be shared and communicated? What sort of channels could be used? For example, radio, video, an existing mobile agro-advisory service; a communication channel targeting children such as a comic, other distribution of channels or agricultural shows.

In most countries there are many different initiatives working to get information to farmers and other stakeholders. It makes sense to find ways to use limited funding wisely and where possible work with partners. Think through what needs to be done (activities) and about the communication materials (posters, videos, pamphlets, radio script, a set of SMS messages etc.) that are needed.

You need to balance using the media that will reach the identified target groups with the resources you have.

Step 9: Which partners will you need to implement the campaign?

The campaign strategy will outline the activities or set of activities required to meet the set campaign objectives. These could be single or a combination of activities, one-off, a series, or repeated activities (e.g. radio programs, SMS campaign, field days and demos on the ISFM technologies, etc.).

There is need to think about how these activities will be implemented - how can they be embedded into the partners work plans? Should some be implemented as additional stand-alone activities?

Some times this will involve contracts or memorandum of understanding - other times you will be sharing objectives and may work together very informally. It is important to clarify what you expect of any person or organization to contribute to the campaign.

Step 10: Do your plans fit your budget?

Stakeholders who will participate in the dissemination or scale-up campaign and the required resources - see step 2. Now is the time to double-check.

Step 11: What does success look like?

There is need to think about how the scale-up campaign will be monitored and evaluated.

Some campaigns include an option the opportunity for tracking results - for example sign up to a SMS service or visits to a website. In other cases you may need to be cleaver about how you gather your information - for example if you are promoting a new seed variety can you work with the input dealers to see how much they have sold?

Using real data is cheaper, easier and more reliable than survey data. Some changes can also be easily observed - like moving may be observable

Task: find a couple of simple indicators of success and clarify how data is going to be collected and read any useful M&E reports from similar projects - but remember they are rarely critical or peer reviewed - see what they learned and also what they measured.

Step 12: Work plan and timescale

The campaign strategy document will inform stakeholders about the ISFM information materials being developed, the purpose of the materials i.e. to create awareness about the ISFM technologies being up-scaled and encourage farmers to adopt them. It will define the ASHC project, describe its products and services (L3 materials), how they will be disseminated (campaign objectives) and identify the target users.

The campaign plan is the practical application of the strategy. It will provide details of the necessary actions to achieve one or more campaign objectives. A good campaign plan summarizes **who, what, where, when, and how much** questions for a specified period. In the case of the campaign to promote priority ISFM technologies, the information collated during the write-shop (1-11 above) should answer the following questions:

- What are the ISFM technologies we are up-scaling?
- What uniqueness do the selected technologies have?
- Who are our target users?
- What campaign activities shall be implemented?
- Who will take responsibility for the activities? Who will be the other partners?
- Where will the campaign activities be implemented?
- When will the campaign activities be implemented? How often will they be implemented? How long will the campaign last?
- How much shall be spent on the campaign? (i.e. What resources?)
- How will the campaign be monitored and evaluated? And by who?

The campaign plan will outline the objectives, activity or set of activities required to meet each objective, the target users, time-frame, responsible or lead partner/other partners, budget and means of evaluating activity and indicators of progress.

But it should be short and easily to look up information you need. Short sentences; lots of headings and tables and checklists wherever possible and keep it up to date. It should be a dynamic document.

This could be presented in a matrix form as shown in the hypothetical example below:

Draft Campaign Plan - example

Objectives	Activity	Who?	Timeframe	Responsibility/ Assignment	Budget	Means of verification & Indicators of progress
To create awareness about priority ISFM technologies (name) to farmers in at least 50 districts of central Ghana	Identify champions to lead campaign	All partners-feed information to XXXX	April 2013	ISFM Inc. and Ministry of Agriculture (Kumasi)	No cost	At least 2 visits made to each target district and 2,000 farmers sensitized per year; MoV: Reports on visits
	Official Launch of campaign during field day					
	Produce flyer					
	Produce radio programs on the ISFM technologies					
	Produce videos on the ISFM technologies					
	Hold field days/demos on the technologies/ disseminate flyers on the same					

Bibliography

Adegeye, A.J. and Dittoh, J. S. (1985). *Essentials of Agricultural Economics.* Impact Publishers, Ibadan, Nigeria.

Anaman, K.A. (1988). *African Farm Management: Principles and Application with Examples*. Ghana Universities Press, Accra.

Baker, G. A., O. Grunewald and W. D. Gorman (2002). *Introduction to Food and Agribusiness Management* Prentice Hall.

Barry, P. J., Ellinger, P. N., Hopkin, J. A., & Baker, C. B. (2000). *Financial Management in Agriculture*. Danville, IL: Interstate Publishers, Inc.

Bationo, A., Fairhurst, T., Giller, K., Kelly, V., Lundaka, R., Mando, A., Mapfumo, P., Odour, G., Romnie, D., Vanluuwe, B., Wairegi, Land Zingore, S.(2012). Handbook for Integrated Soil Fertility Management. Thomas Fairhurst (Ed) African Soil Health Consortium, CAB International.

Bationo, A., Waswa, B., Kihara, J., Adolwa, I., Vanluuwe, B. and Saidou, K. (Eds) (2012). *Lessons Learned from Long-term Soil Fertility Management Experiments in Africa*. Springer. Dordrecht/Heidelberg/New York/ London

Bationo, A., Waswa, B., Okeyo, J. M., Maina, F., Kihara, J. and Mokwunye, U (Eds) (2011). *Fighting Poverty in Sub-saharan* Africa: The Multiple Roles of Legumes in Integrated Soil Fertility Management. Springer. Dordrecht/Heidelberg/New York/London

Beierlein, J.G., K.C. Schneeberger and D.D Osburn (2008). *Principles of Agribusiness Management*. Fourth Edition. Long Grove. IL: Waveland Press, Inc.

Boardamn A. E., D.H. Greenberg, A.R. Vinning and D. L. Weimer (2011). *Cost-Benefit Analysis - Concepts and Practice* (Fourth Edition) Prentice Hall.

Brady, N.C. and Weil, R.R. (1999). *The Nature and Properties of Soils.* 12th edition. Prentice - Hall, Inc. New Jersey.

California Fertilizer Associations (2011). *Western Fertilizer Handbook* (9th Edition).Thomson publications.

Defoer, T., Budelman, A., Toulmin, C. & Carter, S.E. eds. (2000). Building common knowledge: participatory learning and action research (part 1). *In* T. Defoer & A. Budelman, eds. *Managing soil fertility in the tropics. A resource guide for participatory learning and action research.* Royal Tropical Institute, Amsterdam.

Dembele, I., Kone, D., Soumare, A., Coulibaly, D., Kone, Y., Ly, B. & Kater, L. (2000). Fallows and field systems in dryland Mali. *In* T. Hilhorst & F. Muchena, eds. *Nutrients on the move. Soil Fertility Dynamics in African Farming Systems,* p. 83-101. London, IIED.

Doran J.W., M. Sarrantonio and M.Liebig (1996). *Soil Health and Sustainability. Advances in Agronomy.* Academy Press, San Diego CA

Doran J.W. and Zeiss M.R. (2000). *Soil Health and Sustainability: Managing the biotic components of soil quality.* University of Nebraska, Lincoln U.S.A.

Gregory, P.J. and Nortcliff, S. (Eds.) (2012). *Soil Conditions and Plant Growth.* Wiley/Blackwell, Oxford.

International Fertilizer Development Center (IFDC) www.ifdc.org

Joshi, A.L., Doyle, P.T. & Oosting, S.J. (1994) *Variation in the Quantity and Quality of fibrous crop residues.* Proc. National Seminar, BAIF Development Research Foundation, Pune, Maharashtra, India, 8-9 February 1994. Pune, India. 174 pp.

Kang, B.T., Wilson, G.F. and Lawson, T.L. (1984). *Alley Cropping, A Stable Alternative to Shifting Cultivation.* International Institute of Tropical Agriculture, Ibadan, Nigeria.

Karlen D.L., Andrews S.S. and Doran J.W. (2001). *Soil Health Current Concepts and Application.,* U.S.S.D. - A.R.S. National Soil Science Laboratory, Ames, Iowa U.S.A.

Kater, L. (2000). *Fallows and Field systems in Dryland Mail. In* T. Milhorst & F. Machena, eds. Nutrients on the move. Soil Fertility Dynamics in African Farming Systems. International Institute for Environment and Development, London. 146 pp.

Kay, R.D., William, M.E. and Duffy, P. A. (2008). *Farm Management* Sixth Edith. McGraw-Hill Series in Agricultural Economics. McGraw-Hill Inc.

Kotter, J. P. (1996). *Leading Change.* Boston: Harvard Business School Press.

Magdoff. F. and Weil. R.R. (eds). (2004). Soil Organic Matter in Sustainable Agriculture. *Advances in Agroecology series*: Vol. 11, CRC Press.

Mengel, K. and Kirkby, E.A. (2001). *Principles of Plant Nutrition*. Springer, Dordrecht/Heidelberg

Plaster, X.P., and Letey, J. (2000). Chapters 10-12. In: *Soil Science and Management* (4th Editions). Delmer Thomson Learning

Powell, J.M. & Williams, T.O. (1993). *Livestock, nutrient cycling and sustainable agriculture in West Africa.* Gatekeeper series No. 37. London, IIED. 15 pp

Powell, M. (1986). Manure for cropping: a case study from central Nigeria. *Expl. Agriculture,* 22: 15-24.

Reijntjes, C., Haverkort, B. & Waters-Bayer, A. (1992). *Farming for the future. An introduction to low-external input and sustainable agriculture.* Leusden, the Netherlands, MacMillan, ILEIA. 250 pp.

Renard, C. ed. (1997). *Crop residues in sustainable mixed crop/livestock farming systems.* New York, CAB International. 322 pp.

Sanginga, N. and Woomer, P.L. (2009). A Manual on Integrated Soil Fertility Management in Africa. *African Crop Science Conference Proceedings,* 9: 357-363.

Van Der Pol, F. (1992) *Soil Mining: an Unseen Contributor to Farm Income in Southern Mali.* Bull. 325. Amsterdam, Royal Tropical Institute. 48 pp.

van Veldhuizen, L., Waters-Bayer, A. and de Zeeuw, H. (1997). *Developing Technology with Farmers: A trainer's guide for participatory learning.* London and New York: Zed Books.

Walaga, C., Egulu, B., Bekunda, M. and Ebanyat, P. (2000). Impact of policy change on soil fertility management in Uganda. *In* T. Hilhorst & F. Muchena, eds. *Nutrients on the move. Soil fertility dynamics in African farming systems.* London, International Institute for Environment and Development.146 pp.

Walker R. W. and Skogerboe, G. V. (1987). Surface Irrigation Theory and Practice: Prentice-Hall, Englewood Cliffs, NJ, U.S.A., Hardcover.

Wiswall, R (2009). The Organic Farmer's Business Handbook: A Complete Guide to Managing Finances, Crops, and Staff - and Making a Profit. Chelsea Green. USA

http://www.climate-airwaves.net/

http://www.aglearn.net/resources/isfm/

http://www.ciat.cgiar.org/work/Africa/Documents/highlight10.pdf

Network for Sustainable Agriculture (agLearn.net) 2004. Bangkok. www.aglearn.net

Colour Plates

Chapter 1: Soil Health and Soil Fertility

Fig. 1. Map of Ghana, Showing the various agro-ecological zones (Page no 7)

Chapter 2: Evaluating Soil Fertility

Chlorosis: Nitrogen (N) deficiency in maize (Page no 20)

Interveinal chlorosis: Iron (Fe) deficiency in maize (Page no. 20)

Necrosis or firing in maize (Page no. 21)

Chapter 3: Water Management

Bare Soil (Page no. 45)

Mulched Soil (Page no. 45)

Building soil conservation bunds (Page no. 45)

A terraced field (Page no. 47)

Land ploughed along the contours (Page no. 48)

Chapter 4: Approaches to Soil Fertility Management

Un-inoculated Soybean   Inoculated Soybean (Page no. 60)

Soybean to be replaced with maize next season (Page no. 64)   Maize to be replaced with soybean next season (Page no. 64)

Chapter 5: The Concept of Integrated Soil Fertility Management (ISFM)

Good soil Good harvest
Better life

Get healthier crops with ISFM
ISFM = Improved Seed + Fertilizer + Manure

Improved Seed

Use improved seed, stem cuttings, suckers and roots because they mature faster, are resistant to weed and disease attack, and are tolerant to drought. Select seeds that work well in your area and combine with fertilizer and organic matter.

Fertilizer

Applying the correct fertilizer in the right amounts, in the right way at the right time will increase the quantity and quality of your yield - especially when combined with organic matter. Using less than the recommended amounts or no fertilizer at all can result in poor yields.

Manure & other organic matter

To make fertilizer work better, add manure or other organic matter that is available in your area. (e.g. chicken dropping, cow dung, stover and compost).

For good yields combine these with good agricultural practices (correct planting time and spacing, timely weeding, not planting the same crop every season and water management)

Africa Soil Health Consortium
CABI, ICRAF Complex. P.O. Box 633-00621 Nairobi, Kenya
(t) +254-20-722 4450 (e) africa@cabi.org www.cabi.org/ashc

(Page no. 68)

Severe erosion on a tractor ploughed field. The gullies have developed as a result of erosion. (Page no. 78)

Mulching (Page no. 79)

Use of Herbicide (Page no. 79)

Chapter 6: Organic Resource Management within ISFM

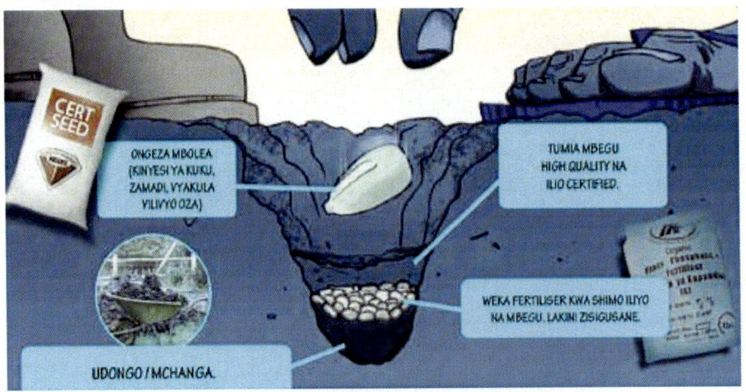

Manure can be microdosed (Page no. 88)

Fertilizer Microdosing Boosting Production in Unproductive Lands (Page no. 89)

NIPA Publications on Agronomy and Soil Science

S.No.	Title	Author	ISBN	Year
1	A Handbook of Minerals,Crystals,Rocks and Ores	Alexander, P.O.	9788190723787	2009
2	A Handbook of Soil-Plant-Water-Fertilizer and Manure Analysis	Durai, M.V.	9789381450185	2014
3	Acid Soils: Their Chemistry and Management	Sarkar, A.K.	9789381450383	2013
4	Advances in Soil Borne Plant Diseases	Naik, Manjunath	9788189422813	2008
5	Agriculture and Waste Management for Sustainable Future	Sannigrahi, A. Kumar	9789380235530	2011
6	Agrometeorology: A Simplified Textbook	Kathiresan, G.	9789383305568	2015
7	Approaches for Incorporating Drought and Salinity Resistance in Crop Plants	Chopra,V.L. & R.S.Paroda	9789383305742	2015
8	Bioinoculants: A Step Towards Sustainable Agriculture	Gupta, R.P. & A.Kalia: et.al.	9788189422219	2007
9	Biotechnology in India: Initiatives and Accomplishments	Niladri, Bag	9789385516252	2016
10	Biotechnology of VA Mycorrhizza: Indian Scenario	Chandra, S. & H.K.Kehri	9788189422226	2006
11	Botanicals as Ecofriendly Pesticides	Mahulikar, P.P. & K.M.Chavan	9788189422578	2007
12	Breeding,Biotechnology and Seed Production of Field Crops	Roy, Bidhan	9789381450680	2013
13	Cereal Grains: Evaluation,Value Addition and Quality Management	Saxena, D.C.:Ed.	9789381450857	2013
14	Cereals: Processing and Nutritional Quality	Ram, Sewa and B.Mishra	9789380235073	2010
15	Climate Change and Agricultural Food Production	Kibria, Golam et.al.	9789381450512	2013
16	Climate Change and Chemicals: Environmental & Biological Aspects	Kibria, Golam et.al.	9789380235301	2010
17	Climate Change and Environment: Concepts and Strategies to Mitigate Impacts	Devesh Sharma & K C Saha	9789385516375	2016
18	Climate Change and Food Security	Datta, M.& N.P.Singh	9788189422387	2008
19	Climate Change and Natural Resources Management	Lenka, S.&N.K.Lenka	9789381450673	2013
20	Climate Change and Plantations in the Humid Tropics	GSHLV Prasada Rao and C.S. Gopakumar	9789385516368	2016
21	Climate Change and Water Security: Impacts, Future Scenarios, Adaptations and Mitigations	Golam Kibria, A. K. Yousuf Haroon & Dayanthi Nugegoda	9789385516269	2016
22	Climate Mitigation and Carbon Finance: Global Initiatives & Challenges	Sahoo, A.K.	9789381450024	2012
23	Conducting An Effective and Successful Training Programme	Sontakki, Bharat S.	9789383305223	2015
24	Conservation Agriculture for Carbon Sequestration and Sustaining Soil Health	Somasundaram, J., R.S.Chaudhary: et.al.	9789383305322	2014
25	Computers in Agriculture	Manish Kumar Sharma et al.	9789385516160	2017
26	Droughts in Agricultural Production: Monitoring & Management	Rao, G.G.S.N.	9789385516009	2015

For details, please visit us at www.nipabooks.com

S.No.	Title	Author	ISBN	Year
27	Encyclopedia of Agricultural Meteorology	Dhaliwal, L.K. & S.S.Hundal	9789380235547	2011
28	Engineering and General Geology	Sawant, P.T.	9789380235516	2011
29	Enhancing Pulses Production: Technologies and Strategies	Gangwar, B. & A.K.Singh	9789381450772	2014
30	Extension Management Strategies for Sustainable Agriculture: Opportunities and Challenges	Philip, T. Rathakrishnan	9789385516313	2016
31	Farm Tools and Equipment for Agriculture	Singh, Surendra	9789385516221	2016
32	Food and Nutritonal Security by Sustainable Agriculture: Methods to Attain and Sustain	Mishra, Kumar Bijesh	9789383305049	2014
33	Geographic Information System	Gurugnanam, B.	9788190851282	2009
34	Geoinformatics Applications in Agriculture	Singh, Anil Kumar & U.K.Chopra	9788189422233	2008
35	Geospatial Technologies for Natural Resources Management	Soam,S.K. & P.D. Sreekant	9789381450802	2013
36	GIS: Fundamentals,Applications and Implementations	Elangovan, K.	9788189422165	2006
37	Good Management Practices for Horticultural Crops	M.K. Jatav, et al.	9789385516641	2016
38	Green Agriculture: Newer Technologies	Behera, K.K.	9789381450277	2012
39	Hill Agriculture: Economics and Sustainability	Sharma, Pawan & Sudhakar Dwivedi	9789381450871	2014
40	Illustrated Dictionary of Microbiology	Patel, Sunil	9788189422950	2008
41	Improving Productivity of Drylands by Sustainable Resource Utilisation and Management	Dayal, Devi	9789385516191	2016
42	Integrated Farming System Practices: Challenges and Opportunities	Nanda, Sankarsana	9789385516207	2016
43	Microbes for Plant Stress Management	D. Joseph Bagyaraj & Jamaluddin	9789385516658	2017
44	Modern Technologies for Sustainable Agriculture	Kumar, Sunil	9789381450611	2013
45	Nanotechnology in Soil Science and Plant Nutrition	Adhikari, Tapan	9789381450789	2013
46	Nitrogen Use Efficiency in Plants	Jain, Vanitha & P. Ananda Kumar	9789380235738	2011
47	Objective Agronomy	Vishwakarma,A.	9789380235127	2010
48	Organic Farming	Singh, A.K.	9789385516139	2015
49	Organic Farming: Scope and Uses of Biofertilizers	Panwar, JDS & Amit Kumar Jain	9789385516184	2016
50	Organic Spices	Parthasarathy, V.A	9788189422844	2008
51	Pest Management and Residual Analysis in Horticultural Crops	Gulati, Rachna & Beena Kumari	9789381450710	2013
52	Pesticides: Methods of Their Residues Estimation	Kumari, Beena & T.S.Kathpal	9789380235394	2010
53	Pigeonpea Hybrids and Their Production	Tikle, A.N.	9789383305957	2015
54	Plant-Microbe Interactions	Ramasamy, K. and K.Kumar	9789383305834	2015
55	Practical Manual of Entomology (Insects and Non-Insects Pests)	Devasahayam, H.Lewin	9789380235905	2011
56	Precision Farming in Horticulture	Singh, Jitendar S.K.Jain,L.K.Dashora	9789381450475	2013
57	Principles and Applications of Agricultural Meteorology	Patra, Alok Kumar	9789385516245	2016
58	Recent Advances in Biopesticides	Johri, Jayendra: eds.	9789380235219	2010
59	Remote Sensing Applications in Dryland Natural Resource Management	Gaur, Mahesh	9789381450321	2013

For details, please visit us at www.nipabooks.com

S.No.	Title	Author	ISBN	Year
60	Samekit Krishi Pranali	Kumar, Sanjeev	9789381450154	2012
61	Seed Production of Field Crops	Mondal, S.S.	9788190723763	2009
62	Seed Science and Technology	Vanangamudi, K.	9789383305117	2014
63	Soil Conservation: Fully Revised and Updated: 3rd ed.	Hudson, Norman	9789383305971	2015
64	Soil Microbiology and Biochemistry	Hassan, G.Dar	9789380235134	2010
65	Soil Sampling and Methods of Analysis	Pal, Sushant	9789381450574	2013
66	Soil Science: An Elementary Textbook	Puri, A.N.	9789383305063	2015
67	Soil Testing and Analysis: Plant, Water and Pesticide Residues	Brajendra, Patiram	9788189422707	2007
68	Solving the Pulses Crisis	Singh, Anil & B.Gangwar	9789381450482	2013
69	Sustainability of Small Farms in Coastal Ecosystems	Satapathy, C	9789385516276	2016
70	Sustainable Agriculture: A Vision for Future	Desai, B.K. and B.T.Pujari	9788189422639	2007
71	Sustainable Environmental Science	Sahu, D.D.	9789381450208	2012
72	System Based Integrated Nutrient Management	Gangwar, B. & V.K.Singh	9789381450055	2012
73	Technologies for Sustainable Green Environment	Davamani, V.	9789381450420	2012
74	The Chemistry of Soil Constituents	Greenland, D.J.	9789385516214	2016
75	The Chemistry of Soil Processes	Greenland, D.J.	9789383305926	2015
76	The Weeds of Kumaun Himalayan Region (Uttarakhand)	T.S. Rana & Bhaskar Datt	9789385516474	2016
77	Weed Science	Das, P.C.	9789383305261	2015
78	Women in Sustainable Agriculture	C.Satapathy & Sabita Mishra	9789383305056	2014